TABLES FOR THE COMPRESSIBLE
FLOW OF DRY AIR

Tables for the Compresible Flow of Dry Air

GIVING

MAJOR PARAMETERS FOR
THE MACH NUMBER RANGE 0 TO 4
$$\Upsilon = 1.403$$

E. L. HOUGHTON
B.Sc.(Eng.). C.Eng., M.I.Mech.E., M.R.Ae.S.
Head of Mechanical and Aeronautical Engineering
The Hatfield Polytchnic

AND

A. E. BROCK
B.Sc.(Eng.). C.Eng., M.R.Ae.S., M.B.C.S.
International Computers Limited

ELSEVIER
BUTTERWORTH
HEINEMANN

AMSTERDAM • BOSTON • HEIDELBERG • LONDON • NEW YORK • OXFORD
PARIS • SAN DIEGO • SAN FRANCISCO • SINGAPORE • SYDNEY • TOKYO

Elsevier Butterworth-Heinemann
Linacre House, Jordan Hill, Oxford OX2 8DP
200 Wheeler Road, Burlington, MA 01803

First published by Arnold 1961
Third edition 1975
Reprinted by Butterworth-Heinemann 2001
Transfered to digital printing 2004

British Library Cataloguing in Publication Data
A catalogue record for this book is available from the British Library

ISBN 0 7131 3352 X

For information on all Elsevier Butterworth-Heinemann
publications visit our website at www.bh.com

Printed and bound by Antony Rowe Ltd, Eastbourne

PREFACE

The aim of this book is to make available a low-priced collection of tables that the student may use throughout his studies and retain for use in his subsequent career as a practising engineer, and to this end the authors have produced an entirely new set of tables for the parameters of Isentropic, Prandtl-Meyer expansive, Rayleigh, Fanno, plane normal and plane oblique shock flows. Constant duct area parameters have also been included, for colleagues concerned with applied thermodynamics and propulsion, working on such aspects as combustion chambers.

The second edition provided an opportunity to present all of the parameters in the tables in a non-dimensional form, so that they can be used in whatever system is convenient.

In this third edition the authors include an additional table of Isentropic flow. Graphical use of the 'method of characteristics' analysis and synthesis of supersonic flow exploits the Prandtl-Meyer expansive flow data when it is most conveniently arranged in terms of equal increments of the flow deflexion angle (from the direction of flow at sonic speed). It is hoped that the range of application of this data is thereby significantly improved.

E. L. HOUGHTON
A. E. BROCK

CONTENTS

NOTE. In these tables, $\gamma = 1·403$.

CONTENTS

SYMBOLS AND NOTATION

A stream tube or duct cross-section area.

A $(M_1{}^2 - 1)$, see equation 96.

a local speed of sound.

B $\frac{1}{4}(\gamma + 1)M_1{}^4 \tan \delta$, see equation 96.

C $\{\frac{1}{2}(\gamma + 1)M_1{}^2 + 1\} \tan \delta$, see equation 96.

c_p specific heat at constant pressure.

c_v specific heat at constant volume.

c ultimate gas velocity.

d duct or stream tube diameter.

J Joule's equivalent.

M Mach number.

$\dot{m}$ mass rate of flow.

n radial flow component in Prandtl-Meyer Expansion.

n normal flow component in plane oblique shock theory.

p air pressure.

q heat exchanged via walls of duct or stream-tube.

ΔS increase of entropy.

s peripheral area of a duct or stream tube.

T Thermodynamic temperature K.

t tangential flow component in Prandtl-Meyer flow.

t tangential flow component in plane oblique shock theory.

u speed of gas flow.

x displacement along axis of duct or stream tube.

x $\cot \beta$, see equation 96.

β shock wave angle, see Fig: 5.

γ ratio of specific heats $= 1 \cdot 403$ for dry air.

Δ semi-vertex angle of a wedge.

δ flow deflexion angle through a plane oblique shock wave, see Fig. 5.

ρ air density.

μ Mach angle $=$ arc sin $(1/M)$.

ν flow deflexion angle in isentropic flow through Prandtl-Meyer expansion, see Fig. 2.

θ angle between local Mach line and the Mach line at $M = 1$, see Fig. 2.

τ_0 surface friction per unit area.

superscript

$*$ denotes quantity when local Mach number equals unity.

suffices

1   denotes quantity ahead of shock wave or general station.

2   denotes quantity behind shock wave or general station.

0   denotes quantity when local gas speed is zero, i.e. at stagnation.

m denotes quantity when the flow deflexion angle through a plane oblique shock wave has its maximum value.

D denotes quantity at some datum or reference station.

(ix)

1. BASIC FLOW EQUATIONS

Consider the steady flow of air in an element of a horizontal stream-tube between sections δx apart where the flow parameters are respectively denoted by suffices 1 and 2, Fig. 1.

If the axis of the tube has little or no curvature and the tube walls are impervious to fluid, and assuming that any integrating processes are confined to flow regimes where $c_p/c_v = \gamma$ is constant (1·403 for air), in the absence of external and body forces, the motion of the element may be examined.

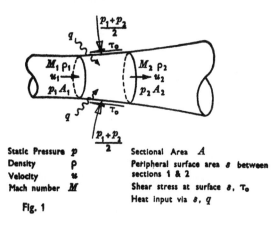

Static Pressure p
Density ρ
Velocity u
Mach number M

Sectional Area A
Peripheral surface area s between sections 1 & 2
Shear stress at surface s, τ_0
Heat input via s, q

Fig. 1

Taking into account:

(i) the potential pressures, which may be assumed to be the mean of the static pressures at the end sections, on the peripheral surface area s,

(ii) the viscous tractions resisting flow at the peripheral surface,

(iii) the heat input q to the element via the peripheral surface,

and invoking the equation of state and equations of conservation of mass, momentum and energy, the following expressions may be derived.

1,1. Equation of State. Making the further assumption that the air obeys the perfect gas law everywhere in the flow

$$p = \rho RT \qquad \ldots \ldots \quad (1)$$

(where R is the characteristic gas constant, the ratio of the universal gas constant and mean molecular weight of the gas) then for the two sections considered

$$\frac{p_1}{\rho_1 T_1} = \frac{p_2}{\rho_2 T_2} \qquad \cdots \qquad \text{(2)}$$

or in terms of the change in state between sections (1) and (2)

$$\delta\left(\frac{p}{\rho T}\right) = 0 \qquad \cdots \qquad \text{(3)}$$

1,2. Conservation of Mass. This is the concept usually expressed in an equation of continuity. In this case, bearing in mind the impervious nature of the walls and the homogeneity of the fluid, the equation is

$$\text{mass flow } \dot{m} = \rho_1 A_1 u_1 = \rho_2 A_2 u_2 \qquad \cdots \qquad \text{(4)}$$

or in terms of the change in mass between sections (1) and (2)

$$\delta(\rho A u) = 0 \qquad \cdots \qquad \text{(5)}$$

1,3. Conservation of Momentum. Since the axis of the tube element is linear the 'directional' sense of the vectors considered may be ignored, and the conservation equation is obtained by equating the rate of change of momentum of the gas in the element to the algebraic sum of all the axial components of force acting on the element.

Thus taking the flow direction as positive:

$$(\rho_1 u_1^2 A_1 - \rho_2 u_2^2 A_2) + (p_1 A_1 - p_2 A_2) - \tau_0 s + \tfrac{1}{2}(p_1 + p_2)(A_2 - A_1) = 0 \qquad \text{(6)}$$

or in terms of changes in forces and momentum rates of change

$$-\delta(\rho u^2 A) - \delta(pA) - \tau_0 s + p\delta A = 0 \qquad \cdots \qquad \text{(7)}$$

1,4. Conservation of Energy. The energy of the gas flowing through section (1) is the sum of its enthalpy and kinetic energy.

The former, h Joules per kilogramme (say), is the sum of the internal energy ($c_v T$) and pressure energy (capacity for work) p/ρ. Thus for unit mass

$$h = c_v T + \frac{p}{\rho} = c_p T$$

The kinetic energy per unit mass of gas flowing is $u^2/2$ so that the total energy flux of the gas per unit mass at section (1) is

$$h_1 = \left(c_p T_1 + \frac{u_1{}^2}{2} \right).$$

This quantity is sometimes referred to as the *stagnation enthalpy*, $H = c_p T_0$.

At section (2) the stagnation enthalpy per unit mass becomes

$$\left(c_p T_2 + \frac{u_2{}^2}{2} \right)$$

and this has been supplemented by the addition of heat q per unit mass to the flow via the walls.

Thus for conservation of energy

$$c_p T_2 + \frac{u_2{}^2}{2} = c_p T_1 + \frac{u_1{}^2}{2} + q \qquad \dots \quad (8)$$

or

$$\delta(c_p T) + \delta\left(\frac{u^2}{2} \right) = q \qquad \dots \quad (9)$$

i.e. $\quad c_p\, \delta T + u\, \delta u = q \qquad \dots \quad (9A)$

Note. It is assumed in this equation that the heat exchange is instantaneous and comprehensive; that is that the heat flows into every part of the gas at the section considered.

1,5. Entropy. Before proceeding to manipulate the equations above it should be borne in mind that the Second Law of thermodynamics must be obeyed in the flow. This introduces a limitation to the flow processes involved which will emerge in later treatment. The Second Law states that the entropy S of a system must increase in a real (irreversible) process and is constant only in an ideal (reversible) process. This can be recognised in the characteristic equation

$$\Delta S \geqslant 0 \qquad \dots \quad (10)$$

The entropy gain from some datum condition (D) to conditions at section (1) is

$$\Delta S_1 = c_p \log \frac{T_1}{T_D} + R \log \frac{p_D}{p_1}.$$

The entropy gain from the datum conditions to conditions at section (2) is

$$\Delta S_2 = c_p \log \frac{T_2}{T_D} + R \log \frac{p_D}{p_2}.$$

Thus equation (10) becomes, from $\Delta S_2 \geqslant \Delta S_1$,

$$c_p \log \frac{T_2}{T_1} + R \log \frac{p_1}{p_2} \geqslant 0 \quad \cdots \quad (11)$$

1,6. The Combined Flow Equation. From the equation of state (3)

$$\frac{p}{\rho T} = \text{constant}$$

or $\log p - \log \rho - \log T = \log (\text{constant}).$

Differentiating, $\dfrac{1}{p} \cdot \dfrac{dp}{d\rho} - \dfrac{1}{\rho} - \dfrac{1}{T} \cdot \dfrac{dT}{d\rho} = 0$

and rearranging

$$\frac{dp}{p} = \frac{d\rho}{\rho} + \frac{dT}{T} \quad \cdots \quad (12)$$

In a similar manner, from considerations of continuity

$$\frac{d\rho}{\rho} + \frac{du}{u} + \frac{dA}{A} = 0 \quad \cdots \quad (13)$$

Eliminating $d\rho/\rho$ from equations (12) and (13) gives

$$\frac{du}{u} + \frac{dA}{A} + \frac{dp}{p} - \frac{dT}{T} = 0 \quad \cdots \quad (14)$$

Now from momentum considerations (equation 7) dp/p can be isolated:

$$- \rho A u \, du - p \, dA - A \, dp + p \, dA - \tau_0 s = 0$$

and dividing by $\rho A u^2$,

$$\frac{du}{u} + \frac{dp}{\rho u^2} + \frac{\tau_0 s}{\rho A u^2} = 0$$

Putting $M^2 = u^2/a^2 = \rho u^2/\gamma p$

$$\frac{du}{u} + \frac{dp}{p} \cdot \frac{1}{\gamma M^2} + \frac{\tau_0 s}{\rho A a^2} \cdot \frac{1}{M^2} = 0$$

and rearranging,

$$\frac{dp}{p} = -\gamma M^2 \frac{du}{u} - \frac{\gamma \tau_0 s}{\rho A a^2} \quad \cdots \quad (15)$$

From energy considerations (equation 9A) dT/T can be isolated :

$$c_p dT = q - u\,du$$

Dividing both sides by $c_p T = a^2/(\gamma - 1)$ gives

$$\frac{dT}{T} = \frac{q}{c_p T} - (\gamma - 1)M^2 \frac{du}{u} \qquad . \quad . \quad . \quad . \quad (16)$$

and substituting the appropriate expressions for dT/T and dp/p in equation (14) gives

$$(M^2 - 1)\frac{du}{u} = \frac{dA}{A} - \frac{q}{c_p T} - \frac{\gamma \tau_0 s}{\rho A a^2} \qquad . \quad . \quad . \quad (17)$$

Equation (17) is useful as a guide to the flow characteristics.

For supersonic flow, i.e. with $M > 1$, it is clear that du will be a negative increment, i.e. the flow will *decelerate*,† if either (a) the tube converges $(dA\,(-\text{ve}))$ or (b) if heat is added $(q\,(+\text{ve}))$ or (c) if friction is present.

Conversely the flow will *accelerate* if (a) the area is increased, i.e. the flow diverges, or (b) heat is withdrawn, or (c) skin friction is reversed. The last condition is unknown, for no mechanism exists for skin friction to act other than in opposition to the flow direction.

For subsonic flow $(M < 1)$ equation (17) can be re-written.

$$(1 - M^2)\frac{du}{u} = -\frac{dA}{A} + \frac{q}{c_p T} + \frac{\gamma \tau_0 s}{\rho A a^2}$$

and flow will *accelerate* $(du\,(+\text{ve}))$ if (a) the flow converges $(dA\,(-\text{ve}))$, (b) if heat is added, (c) if friction is present.

Again conversely the flow will *decelerate* if (a) dA is $(+\text{ve})$ and (b) if heat is withdrawn $(q\,(-\text{ve}))$. Heat may be extracted at the expense of a pressure rise which is in accord with a decelerating subsonic flow regime.

It follows from this brief analysis that considering the effects of area variation, heat addition and friction separately :

(a) Converging the tube will accelerate a subsonic flow to $M =$ unity, and for continued acceleration beyond $M = 1$ the tube must subsequently diverge.

$$\text{i.e.} \quad (1 - M^2)\frac{du}{u} = -\frac{dA}{A}$$

(b) That heat addition will accelerate a subsonic flow, and decelerate a supersonic flow, to $M = $ unity, *and not beyond unity*.

† The existence of the shock wave as the mechanism for decelerating a supersonic flow, a dissipative process accompanied by an entropy increase, has been tacitly ignored here but is considered later in §§6 and 7.

This indicates a limiting quantity of heat, i.e., that sufficient to achieve sonic flow in either case. At this sonic condition the duct is 'choked' and further heat addition reduces the mass flow through the tube or duct.

Heat subtraction has the reverse effect. Withdrawing heat in either flow regime causes the Mach number to recede *from* unity.

(c) Friction effects on the walls of an insulated parallel-sided tube cause a subsonic flow to accelerate to $M = 1$, and a supersonic flow to decelerate to $M = 1$. Again a limiting value for the skin friction appears in that value required to accelerate or decelerate the flow to sonic conditions.

2. ISENTROPIC (FRICTIONLESS ADIABATIC) FLOW

Ignoring skin friction and heat exchange, the equations of state, continuity, conservation of momentum and energy become, respectively, in the notation used in Fig. 1.

$$\frac{p_1}{\rho_1 T_1} = \frac{p_2}{\rho_2 T_2} \quad \text{or} \quad \left(\frac{p}{\rho T}\right) = \text{constant} \qquad . \quad . \quad (18)$$

$$\rho_1 A_1 u_1 = \rho_2 A_2 u_2 \qquad . \quad . \quad . \quad . \quad (19)$$

$$\rho_1 u_1^2 A_1 - \rho_2 u_2^2 A_2 + p_1 A_1 - p_2 A_2 + \tfrac{1}{2}(p_1 + p_2)(A_2 - A_1) = 0 \qquad (20)$$

and
$$c_p T_1 + \tfrac{1}{2} u_1^2 = c_p T_2 + \tfrac{1}{2} u_2^2 \qquad . \quad . \quad . \quad (21)$$

The compressible flow parameter, the *acoustic speed*, can be expressed in several forms:

$$a = \sqrt{\frac{dp}{d\rho}} = \sqrt{\frac{\gamma p}{\rho}} = \sqrt{\gamma R T} = \sqrt{(\gamma - 1) c_p T} = \frac{u}{M}. \quad . \quad (22)$$

Equation (21) can thus be written

$$\left.\begin{array}{c} \dfrac{u_1^2}{2} + \dfrac{\gamma p_1}{(\gamma - 1)\rho_1} = \dfrac{u_2^2}{2} + \dfrac{\gamma p_2}{(\gamma - 1)\rho_2} \\[2mm] \dfrac{u_1^2}{2} + \dfrac{a_1^2}{\gamma - 1} = \dfrac{u_2^2}{2} + \dfrac{a_2^2}{\gamma - 1} \end{array}\right\} \quad . \quad . \quad (23)$$

or

Taking the sections corresponding to (i) the reservoir, where $u = 0$, $M = 0$ and the other quantities are denoted by the suffix 0, (ii) the throat where $u = a$, $M = 1$, and the quantities are denoted by an asterisk, (iii) the ultimate conditions where the flow is expanded to $p = 0$, $T = 0$, and all the energy of the flow is converted to kinetic energy† so that the velocity is the maximum value, c, equations (23) give

$$\underbrace{\frac{\gamma p_0}{(\gamma-1)\rho_0} = c_p T_0 = \frac{a_0{}^2}{\gamma-1}}_{\text{reservoir conditions}} = \underbrace{\frac{\gamma+1}{2(\gamma-1)}\, a^{*2} = \frac{\gamma+1}{2}\, c_p T^*}_{\text{throat}} = \underbrace{\frac{c^2}{2}}_{\text{ultimate}} = \text{etc.} \tag{24}$$

Manipulation of equations (23) between conditions at the reservoir and any section (without suffix) gives:

2,1. Pressure Ratio

$$\frac{p_0}{p} = \left[1+\frac{\gamma-1}{2}\,M^2\right]^{\gamma/(\gamma-1)} \quad \ldots\ldots \tag{25}$$

2,2. Density Ratio

$$\frac{\rho_0}{\rho} = \left[1+\frac{\gamma-1}{2}\,M^2\right]^{1/(\gamma-1)} \quad \ldots\ldots \tag{26}$$

2,3. Temperature Ratio

$$\frac{T_0}{T} = \frac{p_0/p}{\rho_0/\rho} = \left[1+\frac{\gamma-1}{2}\,M^2\right] \quad \ldots\ldots \tag{27}$$

2,4. Area Ratio. Invoking the equation of continuity (19) at these sections

$$\dot{m} = \rho A u = \frac{\rho}{\rho_0}\cdot\rho_0 A u = \rho_0 A u\!\left(\frac{p}{p_0}\right)^{1/\gamma}$$

Manipulating equations (23) via equations (24) to give

$$u = a_0\sqrt{\left[1-\left(\frac{p}{p_0}\right)^{(\gamma-1)/\gamma}\right]\frac{2}{\gamma-1}},$$

and putting $\sqrt{\dfrac{\gamma p_0}{\rho_0}}$ for a_0

$$\frac{\dot{m}}{A} = \left(\frac{p}{p_0}\right)^{1/\gamma}\sqrt{\frac{2\gamma}{\gamma-1}\,p_0\rho_0\left[1-\left(\frac{p}{p_0}\right)^{(\gamma-1)/\gamma}\right]} \quad . \tag{28}$$

Applying the same equation to the reservoir and throat stations gives with

$$\frac{p^*}{p_0} = \left(1+\frac{\gamma-1}{2}\right)^{-\gamma/(\gamma-1)} = \left(\frac{\gamma+1}{2}\right)^{-\gamma/(\gamma-1)} \quad \text{and from equation} \quad (25)$$

† A real gas will liquefy before this state is reached.

$$\frac{\dot{m}}{A^*} = \sqrt{p_0 \rho_0 \gamma \left(\frac{2}{\gamma+1}\right)^{(\gamma+1)/(\gamma-1)}} \quad . \quad . \quad . \quad . \quad (29)$$

Eliminating $\dot{m}$ from equations (28) and (29) and writing p/p_0 in terms of M from equation (25) gives the area ratio

$$\frac{A}{A^*} = \frac{1}{M} \left[\frac{1+\dfrac{\gamma-1}{2}M^2}{\dfrac{\gamma+1}{2}}\right]^{(\gamma+1)/(2(\gamma-1))} \quad . \quad . \quad (30)$$

2,5. Velocity Ratio. The velocity can be directly expressed as a fraction of the ultimate speed it can attain starting from the appropriate reservoir conditions:

$$\frac{u^2}{c^2} = \frac{M^2 a^2}{c^2} = M^2 \frac{(\gamma-1)c_p T}{2c_p T_0}$$

or

$$\frac{u}{c} = M\sqrt{\frac{\gamma-1}{2}}\sqrt{\frac{T}{T_0}} \quad . \quad . \quad . \quad . \quad (31)$$

and substituting for $\sqrt{(T/T_0)}$ from equation (27)

$$\frac{u}{c} = M\sqrt{\frac{\gamma-1}{2+(\gamma-1)M^2}} \quad . \quad . \quad . \quad (32)$$

3. PRANDTL-MEYER EXPANSION

The local flow parameters in the supersonic regime of the flow analysed in the preceding section are identical in magnitude in every respect to the corresponding conditions obtained locally in a supersonic flow which is accelerated by expansion round a corner.

In Fig. 2 the flow is accelerated to a general state by isentropic expansion through a deflexion angle ν from sonic (throat) conditions as datum.

Considerations of continuity and momentum, now of course expanded to accommodate (two-dimensional) flow along curved paths, show that the only practical flow possible is one where:

(a) the flow parameters are constant along radial ordinates, and
(b) the local velocity normal to the radius vector is the local acoustic speed a.

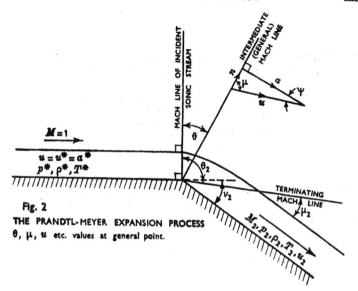

Fig. 2
THE PRANDTL-MEYER EXPANSION PROCESS
θ, μ, u etc. values at general point.

This implies that the radius vector anywhere in the expansion is the Mach line of the local stream and that the flow is inclined at μ to the radius vector, where

$$\mu = \text{arc sin} \frac{1}{M} \qquad \qquad (33)$$

The radius vector moves through angle θ, as the flow deflects through ν which is given by simple geometry as

$$\nu = \theta - \psi \qquad \qquad (34)$$

and $\psi = \text{arc tan } (n/a)$, where n is the radial flow velocity component.

Rewriting the energy equations (23) and (24) with the radial and tangential velocity components introduced, gives

$$\frac{n^2 + a^2}{2} + \frac{a^2}{\gamma - 1} = \frac{c^2}{2}$$

or

$$a = \sqrt{\frac{\gamma - 1}{\gamma + 1} (c^2 - n^2)} \qquad \qquad (35)$$

From introducing condition (a) into the momentum equations, or invoking the existence of a velocity potential (zero vorticity)

$$a = \frac{\partial n}{\partial \theta} \qquad \qquad (36)$$

Combining equations (35) and (36) and integrating from the datum conditions where $a = a^*$, $n = 0$, $\theta = 0$

$$n = c \sin \sqrt{\frac{\gamma-1}{\gamma+1}} \theta \quad \cdots \quad \cdots \quad (37)$$

and
$$a = \sqrt{\frac{\gamma-1}{\gamma+1}} c \cos \sqrt{\frac{\gamma-1}{\gamma+1}} \theta \quad \cdots \quad (38)$$

Then anywhere in the expansion, from equations (37) and (38),

$$M = \sqrt{\frac{n^2+a^2}{a^2}} = \sqrt{1+\frac{\gamma+1}{\gamma-1} \tan^2 \sqrt{\frac{\gamma-1}{\gamma+1}} \theta}$$

which rearranged, gives

$$\theta = \sqrt{\frac{\gamma+1}{\gamma-1}} \text{ arc cos} \sqrt{\frac{\gamma+1}{2+(\gamma-1)M^2}} \quad \cdots \quad (39)$$

$$\psi = \text{arc tan} \frac{n}{a} = \text{arc tan} \left\{ \sqrt{\frac{\gamma+1}{\gamma-1}} \tan \sqrt{\frac{\gamma-1}{\gamma+1}} \theta \right\}$$

$$= \text{arc tan} \sqrt{M^2-1}$$

or from geometry $\quad \psi = \frac{\pi}{2} - \mu = \frac{\pi}{2} - \text{arc sin} \frac{1}{M}.$

Substituting in equation (34) for θ and ψ gives

$$\nu = \sqrt{\frac{\gamma+1}{\gamma-1}} \text{ arc cos} \sqrt{\frac{\gamma+1}{2+(\gamma-1)M^2}} - \text{arc tan} \sqrt{M^2-1} \quad (40)$$

4. FRICTIONLESS FLOW IN A CONSTANT AREA DUCT WITH HEAT EXCHANGE (Rayleigh Flow)

The general equation of §1,6 is now considered with the heat addition to (or subtraction from) the flow between sections as the major physical variant.

A particular assumption made in this case is that the ratio of specific heats remains constant. If the heat addition is acquired by chemical processes (combustion) or if the temperature becomes large, the assumption is in serious error. Alternatively the simple transfer of heat via the walls of the duct is unlikely not to be accompanied by surface friction, a flow characteristic dealt with in §5. The following theory and tables must accordingly be treated with the usual care reserved for idealised predictions.

As shown in §1,6, for either subsonic or supersonic flow, heat *addition* causes the Mach number to *approach* unity, whilst heat *withdrawal* makes the Mach number *recede* from unity.

Sonic or throat conditions then offer a limitation to the heat addition in any particular case *for constant mass flow*. If further heat is added the mass flowing per second is reduced. When the mass flow in the duct attains the critical (sonic) value it is said to be choked.

In a supersonic flow of real fluid, deceleration is accompanied by a system of shock waves, thus the simplifying assumptions are effectively destroyed; equally the heat input may be accompanied by flame phenomena. The ratios obtained by the following simplified treatment are still useful, however, providing precautions are taken.

Note that in this type of flow the stagnation enthalpy and hence ultimate gas velocity are *increased* if heat is added.

4,1. Pressure Ratio. The equations of motion, continuity, momentum and energy are for this case, with $A_1 = A_2 =$ constant, as follows:

$$\frac{p_1}{\rho_1 T_1} = \frac{p_2}{\rho_2 T_2} \qquad \qquad (41)$$

$$\rho_1 u_1 = \rho_2 u_2 \qquad \qquad (42)$$

$$\rho_1 u_1{}^2 - \rho_2 u_2{}^2 + (p_1 - p_2) = 0 \qquad \cdots \quad (43)$$

$$c_p T_2 + \tfrac{1}{2} u_2{}^2 = c_p T_1 + \tfrac{1}{2} u_1{}^2 + q \qquad \cdots \quad (44)$$

From equation (43) $\rho_1 u_1{}^2 + p_1 = \rho_2 u_2{}^2 + p_2$

or $\gamma p_1 M_1{}^2 + p_1 = \gamma p_2 M_2{}^2 + p_2$

and setting conditions (2) to be critical (or sonic)

$$p_1(1 + \gamma M^2) = p^*(1 + \gamma)$$

Therefore $\dfrac{p}{p^*} = \dfrac{1 + \gamma}{1 + \gamma M^2} \qquad \cdots \quad (45)$

4,2. Temperature Ratio. From equation (41)

$$\frac{T}{T^*} = \frac{p \rho^*}{p^* \rho}$$

$$= \frac{p}{p^*} \cdot \frac{u}{u^*}$$

But $u = Ma$ and $u^* = a^*$

$$\frac{T}{T^*} = \frac{p}{p^*} \cdot M \frac{a}{a^*} = \frac{p}{p^*} \cdot M \sqrt{\frac{T}{T^*}}$$

Therefore
$$\frac{T}{T^*} = M^2 \left(\frac{1+\gamma}{1+\gamma M^2} \right)^2 \qquad \cdots \quad (46)$$

4,3. Density Ratio. Directly from equation (41)

$$\frac{\rho}{\rho^*} = \frac{p}{p^*} \cdot \frac{T^*}{T} = \frac{1+\gamma M^2}{(1+\gamma)M^2} \qquad \cdots \quad (47)$$

4,4. Total Pressure Ratio. Defining the total pressure p_0 at any section as the pressure attained by bringing the fluid to rest isentropically, equation (25) gives

$$\frac{p_{0,1}}{p_1} = \left[1 + \frac{\gamma-1}{2} M^2 \right]^{\gamma/(\gamma-1)}$$

For the critical section M = unity and

$$\frac{p_0^*}{p^*} = \left(1 + \frac{\gamma-1}{2} \right)^{\gamma/(\gamma-1)}$$

Thus
$$\frac{p_0}{p_0^*} = \left[\frac{1 + \frac{\gamma-1}{2} M^2}{1 + \frac{\gamma-1}{2}} \right]^{\gamma/(\gamma-1)} \cdot \frac{p}{p^*}$$

and substituting for p/p^* from equation (45)

$$\frac{p_0}{p_0^*} = \left(\frac{1+\gamma}{1+\gamma M^2} \right) \left(\frac{2+(\gamma-1)M^2}{\gamma+1} \right)^{\gamma/(\gamma-1)} \qquad \cdots \quad (48)$$

4,5. Total Temperature Ratio. In a like manner total temperature is the temperature attained by bringing the gas to rest adiabatically, and from equation (27)

$$\frac{T_0}{T} = 1 + \frac{\gamma-1}{2} M^2, \quad \frac{T_0^*}{T^*} = 1 + \frac{\gamma-1}{2}$$

Therefore
$$\frac{T_0}{T_0^*} = \frac{2+(\gamma-1)M^2}{\gamma+1} \cdot \frac{T}{T^*}$$

and substituting for T/T^*

$$\frac{T_0}{T_0^*} = \frac{(1+\gamma)M^2}{(1+\gamma M^2)^2}[2+(\gamma-1)M^2] \qquad \cdots \quad (49)$$

Note. The total temperature as defined above is directly proportional to the stagnation enthalpy of §1,4 so that equations (8) and (44) can be given as

$$c_p T_{0,2} - c_p T_{0,1} = q \quad \ldots \ldots \quad (50)$$

showing that the stagnation enthalpy is increased or decreased by the heat exchange.

4,6. Heat Transfer Ratio. Differentiating equation (49) with respect to M and equating to zero yields a maximum at $M=1$, corroborating §4,0 above.

From equation (50), the heat required to produce critical conditions is $q = (c_p T_0^* - c_p T_0)$, giving

$$\frac{q}{c_p T_0^*} = 1 - \frac{T_0}{T_0^*} = \left(\frac{1+\gamma M^2}{1-M^2}\right)^{-2} \quad \ldots \quad (51)$$

4,7. Velocity Ratio. Finally for this case $u/u^* = \rho^*/\rho$ from equation (42) and on substituting from equation (47)

$$\frac{u}{u^*} = \frac{(1+\gamma)M^2}{1+\gamma M^2} \quad \ldots \ldots \quad (52)$$

4,8. Entropy Change. From equation (11)

$$\frac{\Delta S}{c_v} = \gamma \log \frac{T}{T^*} + (\gamma-1) \log \frac{p^*}{p} \quad \ldots \quad (11A)$$

and substituting the appropriate expressions for T/T^* and p/p^*

$$\frac{\Delta S}{c_v} = \gamma \log \left\{ M^2 \left[\frac{\gamma+1}{1+\gamma M^2}\right]^{(\gamma+1)/\gamma} \right\} \quad \ldots \quad (53)$$

5. ADIABATIC FLOW IN A CONSTANT AREA DUCT WITH SURFACE FRICTION (Fanno Flow)

Again the basic assumptions represent somewhat artificial conditions. At any section in a real gas flow the values of the parameters are far from uniform, e.g. the condition of 'no slip' at the wall causes an extreme variation in the velocity. However, if suitable mean quantities are used little loss in accuracy results. In the supersonic flow case, the existence of a shock system may also destroy the ideal concept of the flow. In fact isentropic pressure recovery as suggested in the theory is impossible. However this model produces typical values for the variants which differ little from mean values experienced under controlled conditions in real pipes.

In the subsonic case, the wall friction increases the velocity and decreases the pressure for a given mass flow until critical values are reached. The critical value of the velocity is the local acoustic speed where M is unity. This case is directly analogous to the reduction in area increasing the velocity to sonic conditions as in §2 and the physical interpretation could well include an effective reduction in area due to the growth of displacement thickness of the boundary layer.

In supersonic pipe flows wall friction reduces the velocity and the pressure rises until sonic conditions are reached. In both flow regimes extension of the pipe length beyond the critical causes a reduction in the mass flow.

In what follows the parameters are expressed as ratios of the critical (or sonic) parameters as before.

The surface friction coefficient C_f, defined as the ratio of wall shear stress to dynamic pressure of local (mean) pipe flow, is a function of Reynolds number, which is usually expressed in terms of the pipe diameter once steady conditions away from the entry are attained. Once again an average value may be used. The surface friction coefficient varies only slightly with M and is usually assumed constant in that respect.

The pipe length over which the conditions in any problem change is introduced in the dimensionless parameter

$$\frac{4C_f}{d}[x^* - x] \text{ which is always} \geqslant 0.$$

It becomes zero at the critical condition (where $M=1$). Since C_f and d are finite, $x=x^*$ for this condition which represents the maximum length of the pipe for the particular inlet conditions considered.

For the case of wall friction, constant area and zero heat exchange, the equations of state, continuity and conservation of momentum and energy become:

$$\frac{p_1}{\rho_1 T_1} = \frac{p_2}{\rho_2 T_2} \qquad \cdots \qquad \cdots \quad (54)$$

$$\rho_1 u_1 = \rho_2 u_2 \qquad \cdots \quad \cdots \quad (55)$$

$$\rho_1 u_1{}^2 - \rho_2 u_2{}^2 - \tau_0 s/A + (p_1 - p_2) = 0 \qquad \cdots \quad (56)$$

$$c_p T_2 + \tfrac{1}{2} u_2{}^2 = c_p T_1 + \tfrac{1}{2} u_1{}^2 \qquad \cdots \quad \cdots \quad (57)$$

5,1. Temperature Ratio. Manipulating equation (57) as in the isentropic flow case, §2, gives, via equations (23) and (24),

$$\frac{T_0}{T} = \left[1 + \frac{\gamma - 1}{2} M^2\right] \qquad \dots \quad (27)$$

or

$$\frac{T_0}{T^*} = \left[1 + \frac{\gamma - 1}{2}\right] = \frac{\gamma + 1}{2}$$

whence

$$\frac{T}{T^*} = \frac{\gamma + 1}{2 + (\gamma - 1)M^2} \qquad \dots \quad (58)$$

5,2. Velocity Ratio

$$M = \frac{u}{a} = \frac{u}{\sqrt{\gamma R T}} = u \Big/ \left[\gamma R T_0 \left(1 + \frac{\gamma - 1}{2} M^2\right)^{-1}\right]^{1/2}$$

which on rearranging gives

$$u = \left[\frac{\gamma R T_0 M^2}{1 + \frac{\gamma - 1}{2} M^2}\right]^{1/2} \qquad \dots \quad (59)$$

In particular, for $M = 1$

$$u^* = \sqrt{\gamma R T_0}\left(\frac{2}{\gamma + 1}\right)^{1/2}$$

giving

$$\frac{u}{u^*} = M \left[\frac{\gamma + 1}{2 + (\gamma - 1)M^2}\right]^{1/2} \qquad \dots \quad (60)$$

5,3. Density Ratio. Directly from continuity equation (55)

$$\frac{\rho}{\rho^*} = \frac{u^*}{u}$$

$$\frac{\rho}{\rho^*} = \left[\frac{2 + (\gamma - 1)M^2}{(\gamma + 1)M^2}\right]^{1/2} \qquad \dots \quad (61)$$

5,4. Static Pressure Ratio. From the equation of state (54) $p/p^* = (\rho/\rho^*)(T/T^*)$

whence

$$\frac{p}{p^*} = \frac{1}{M}\left[\frac{\gamma + 1}{2 + (\gamma - 1)M^2}\right]^{1/2} \qquad \dots \quad (62)$$

5,5. Total Pressure Ratio. Introducing again the total pressure

$$p_0 = p\left[1 + \frac{\gamma - 1}{2} M^2\right]^{\gamma/(\gamma - 1)} \qquad \text{(from equation (25))}$$

$$p_0^* = p^* \left[1 + \frac{\gamma-1}{2} \right]^{\gamma/(\gamma-1)} = p^* \left[\frac{\gamma+1}{2} \right]^{\gamma/(\gamma-1)}$$

giving

$$\frac{p_0}{p_0^*} = \frac{p}{p^*} \left[\frac{2+(\gamma-1)M^2}{\gamma+1} \right]^{\gamma/(\gamma-1)}$$

and on substituting for p/p^* from equation (62)

$$\frac{p_0}{p_0^*} = \frac{1}{M} \left[\frac{2+(\gamma-1)M^2}{\gamma+1} \right]^{\frac{1}{2} \cdot (\gamma+1)/(\gamma-1)} \qquad . \quad . \quad (63)$$

5,6. The Length-Friction Coefficient Parameter. The shear force at the walls of a length δx of a circular duct of area $A \; (= \pi d^2/4)$ is written in coefficient form as $C_f \frac{1}{2} \rho u^2 \pi d \delta x$, and putting these expressions in the appropriate term of equation (15) gives:

$$\frac{dp}{p} + \gamma M^2 \frac{du}{u} = - \frac{\gamma C_f \frac{1}{2} \rho u^2 \pi d \delta x}{\rho \frac{\pi}{4} d^2 a^2}$$

and cancelling and rearranging

$$\frac{1}{\gamma M^2} \cdot \frac{dp}{p} + \frac{du}{u} = \frac{-C_f}{2} \cdot \frac{4}{d} \cdot \delta x \qquad . \quad . \quad . \quad . \quad (64)$$

The ratios in the left-hand side can be expressed in terms of M as follows:

from equation (62) $\quad \dfrac{dp}{p} = \dfrac{d\left(M^2 + \dfrac{\gamma-1}{2} M^4 \right)^{-(1/2)}}{\left(M^2 + \dfrac{\gamma-1}{2} M^4 \right)^{-(1/2)}}$

Differentiating etc.,

$$\frac{1}{\gamma M^2} \frac{dp}{p} = - \frac{dM}{\gamma M^3} - \frac{(\gamma-1)dM}{2\gamma M \left(1 + \dfrac{\gamma-1}{2} M^2 \right)}$$

$$= \frac{d}{dM} \left\{ \frac{1}{2\gamma M^2} + \frac{\gamma-1}{4\gamma} \log \left(\frac{1 + \dfrac{\gamma-1}{2} M^2}{M^2} \right) \right\} dM \qquad . \quad . \quad . \quad (65)$$

Likewise from equation (59)

$$\frac{du}{u} = d \left(\frac{M^2}{1 + \dfrac{\gamma-1}{2} M^2} \right)^{1/2} \Bigg/ \left(\frac{M^2}{1 + \dfrac{\gamma-1}{2} M^2} \right)^{1/2}$$

which reduces to

$$\frac{du}{u} = \frac{dM}{M} - \frac{\gamma-1}{2} \frac{M}{1+\frac{\gamma-1}{2}M^2} dM$$

or

$$\frac{du}{u} = \frac{d}{dM}\left\{-\tfrac{1}{2}\log\left(\frac{1+\frac{\gamma-1}{2}M^2}{M^2}\right)\right\}dM \qquad . \quad . \quad (66)$$

Substituting for the appropriate expressions in M from equations (65) and (66) and integrating both sides of equation (64) along the duct from conditions at M, x, to critical sonic conditions $M=$ unity, $x=x^*$

$$\int_M^1 \frac{1}{M^2}\cdot\frac{dp}{p} + \int_M^1 \frac{du}{u} = -\int_x^{x^*} \frac{C_f}{2}\cdot\frac{4}{d}\,dx$$

$$\left[\frac{1}{2\gamma M^2} + \frac{\gamma-1}{4\gamma}\log\left(\frac{1+\tfrac{1}{2}(\gamma-1)M^2}{M^2}\right) - \frac{1}{2}\log\left(\frac{1+\tfrac{1}{2}(\gamma-1)M^2}{M^2}\right)\right]_M^1$$

$$= \frac{-C_f}{2}\cdot\frac{4}{d}(x-x^*)$$

Reversing the limits

$$-\left[\frac{1}{\gamma M^2} + \frac{\gamma+1}{2\gamma}\log\frac{M^2}{1+\frac{\gamma-1}{2}M^2}\right]_1^M = -C_f\frac{4}{d}(x-x^*)$$

or $\quad -C_f\frac{4}{d}(x-x^*) = \frac{1}{\gamma}\frac{1-M^2}{M^2} + \frac{\gamma+1}{2\gamma}\log\left[\frac{\frac{\gamma+1}{2}M^2}{1+\frac{\gamma-1}{2}M^2}\right]$ (67)

For non-circular ducts an *hydraulic radius* may be used.

5,7. The Entropy Change. From equation (11)

$$\frac{\Delta S}{c_v} = \gamma \log\frac{T}{T^*} + (\gamma-1)\log\frac{p^*}{p} \qquad . \quad . \quad . \quad (11A)$$

and substituting the appropriate expressions for p/p^* and T/T^* gives

$$\frac{\Delta S}{c_v} = \gamma \log\left[M^2\left\{\frac{\gamma+1}{(2+(\gamma-1)M^2)M^2}\right\}^{(\gamma+1)/(2\gamma)}\right] \qquad . \quad . \quad (68)$$

6. PLANE NORMAL SHOCK WAVES

Consider again the frictionless adiabatic flow of a perfect gas in a constant area insulated duct. Since no wall effects are included, the model may well be reduced to the case of a stream tube of unit area in the adiabatic flow of a compressible fluid, Fig. 3.

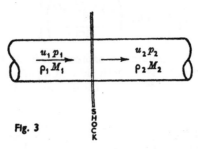

Fig. 3

Equations (83) and (84) below show that the only flow possible in a frictionless tube of constant area, except for the uniform flow case $u_1 = u_2$, $p_1 = p_2$, $\rho_1 = \rho_2$ etc., is a compressive flow from initially *supersonic* conditions to a finally *subsonic* condition. This is always accompanied by an entropy gain. Any form of subsonic flow other than uniform is impossible.

Further, the compression can take place abruptly and the equations are in fact obtained as if a non-impulsive, insulated discontinuity occurs between the sections (1) and (2), the remaining flow being appropriately uniform.

The discontinuity is called a *shock wave*. That it has finite thickness and characteristic properties of its own does not invalidate the following treatment.

The equations appropriate to this flow, rewritten from equations (2), (6) and (8) are

$$\frac{p_1}{\rho_1 T_1} = \frac{p_2}{\rho_2 T_2} \quad\quad\quad\quad (69)$$

$$\dot{m} = \rho_1 u_1 = \rho_2 u_2 \quad\quad\quad (70)$$

$$\rho_1 u_1{}^2 - \rho_2 u_2{}^2 + (p_1 - p_2) = 0 \quad\quad (71)$$

$$c_p T_1 + \tfrac{1}{2} u_1{}^2 = c_p T_2 + \tfrac{1}{2} u_2{}^2 = c_p T_0 \quad\quad (72)$$

6,1. Rankine-Hugoniot Relationships. Writing $\{\gamma/(\gamma-1)\}p/\rho$ for $c_p T$ and rearranging equation (72)

$$\frac{\gamma}{\gamma-1}\left(\frac{p_1}{\rho_1} - \frac{p_2}{\rho_2}\right) = \tfrac{1}{2}(u_2 - u_1)(u_2 + u_1) \quad\quad (73)$$

$$= \tfrac{1}{2}(p_1 - p_2)\left(\frac{1}{\rho_2} + \frac{1}{\rho_1}\right) \quad \cdot \quad \cdot \quad (74)$$

from equations (70) and (71).

Rearranging gives the Rankine-Hugoniot relationships:

$$\frac{p_2}{p_1} = \frac{\dfrac{\gamma+1}{\gamma-1} \cdot \dfrac{\rho_2}{\rho_1} - 1}{\dfrac{\gamma+1}{\gamma-1} - \dfrac{\rho_2}{\rho_1}} \quad \cdots \cdots \quad (75)$$

and
$$\frac{\rho_2}{\rho_1} = \frac{\dfrac{\gamma+1}{\gamma-1} \cdot \dfrac{p_2}{p_1} + 1}{\dfrac{\gamma+1}{\gamma-1} + \dfrac{p_2}{p_1}} \quad \cdots \cdots \quad (76)$$

6,2. Pressure Ratio Across Shock. Again from the momentum equation (71) and the continuity equation (70)

$$\frac{p_2 - p_1}{p_1} = \frac{\rho_1 u_1{}^2 - \rho_2 u_2{}^2}{p_1} = \gamma M_1{}^2\left(1 - \frac{u_2}{u_1}\right)$$

and substituting for ρ_2/ρ_1 ($= u_1/u_2$) from equation (76) and reducing

$$\frac{p_2}{p_1} = \frac{2\gamma}{\gamma+1} M_1{}^2 - \frac{\gamma-1}{\gamma+1} \quad \cdots \quad (77)$$

6,3. Exit Mach Number. By manipulating the variables in the reverse order

$$\frac{p_1}{p_2} = \frac{2\gamma}{\gamma+1} M_2{}^2 - \frac{\gamma-1}{\gamma+1} \quad \cdots \quad (78)$$

Eliminating p_2/p_1 from equations (77) and (78) by simple multiplication gives

$$M_2{}^2 = \frac{(\gamma-1)M_1{}^2 + 2}{2\gamma M_1{}^2 - (\gamma-1)} \quad \cdots \quad (79)$$

6,4. Density Ratio across Shock. Substituting for p_2/p_1 from equation (77) into equation (76) gives

$$\frac{\rho_2}{\rho_1} = \frac{(\gamma+1)M_1{}^2}{2 + (\gamma-1)M_1{}^2} \quad \cdots \quad (80)$$

6,5. Temperature Ratio across Shock.

Since $T_2/T_1 = (p_2/p_1)/(\rho_2/\rho_1)$,

$$\frac{T_2}{T_1} = \left(\frac{2\gamma M_1{}^2 - (\gamma-1)}{\gamma+1}\right)\left(\frac{2 + (\gamma-1)M_1{}^2}{(\gamma+1)M_1{}^2}\right) \quad \cdot \quad \cdot \quad (81)$$

6,6. Entropy Change across Shock. For the entropy change in this flow case equation (11) gives

$$\frac{\Delta S}{c_v} = \gamma \log \frac{T_2}{T_1} + (\gamma - 1) \log \frac{p_1}{p_2} \quad \dots \quad (11B)$$

Substituting for T_2/T_1 and p_1/p_2 from equations (81) and (78)

$$\frac{\Delta S}{c_v} = \log\left\{\left(\frac{2 + (\gamma - 1)M_1{}^2}{(\gamma + 1)M_1{}^2}\right)^{\gamma}\left(\frac{2\gamma M_1{}^2 - (\gamma - 1)}{\gamma + 1}\right)^{\gamma - 1}\right\} \quad (82)$$

which on expanding gives for the first term of a series

$$\frac{\Delta S}{c_v} = \frac{2\gamma(\gamma - 1)}{(\gamma + 1)^2} \cdot \frac{(M_1{}^2 - 1)^3}{3} \quad \dots \quad (83)$$

For $M_1 > 1$, $\Delta S > 0$ and the flow process involves a dissipation. For $M_1 < 1$, $\Delta S < 0$, which is inadmissible by the Second Law of thermodynamics.

From equation (72) on substituting $c_p T_1 = \frac{\gamma}{\gamma - 1} \cdot \frac{p_1}{\rho_1}$

$$\frac{p_1}{\rho_1} = \frac{\gamma - 1}{\gamma}\left\{c_p T_0 - \frac{u_1{}^2}{2}\right\}, \quad \frac{p_2}{\rho_2} = \frac{\gamma - 1}{\gamma}\left\{c_p T_0 - \frac{u_2{}^2}{2}\right\}$$

From equations (71) and (70)

$$u_1 - u_2 = \frac{p_2}{\rho_2 u_2} - \frac{p_1}{\rho_1 u_1}$$

and substituting from the preceding line

$$u_1 - u_2 = \frac{\gamma - 1}{\gamma}\left\{(u_1 - u_2)\left(\frac{1}{2} + \frac{c_p T_0}{u_1 u_2}\right)\right\}$$

Ignoring the trivial solution $u_1 = u_2$

$$u_1 u_2 = \frac{2(\gamma - 1)}{\gamma + 1} c_p T_0 = a^{*2} \quad \dots \quad (84)$$

Note that for the limiting case of sonic flow $M = 1$, $u_1 = u_2 = a^*$.

6,7. The Velocity Change across Shock. From continuity $u_2/u_1 = \rho_1/\rho_2$, so from equation (80)

$$\frac{u_2}{u_1} = \frac{2 + (\gamma - 1)M_1{}^2}{(\gamma + 1)M_1{}^2} \quad \dots \quad (85)$$

6,8. Pressure Recovery. The pressure recovery factor $R_{0,1}$ is defined as $p_{0,2}/p_{0,1}$

Thus

$$R_{0,\,1} = \frac{p_{0,2}}{p_1} \cdot \frac{p_1}{p_{0,1}}$$

Although this factor is not given explicitly in the tables, $p_{0,2}/p_1$ is given in Table IV while $p_{0,1}/p_1$ is given in Table I.

If a tube effectively closed at one end (by a pressure measuring device, say) has its other end open and pointing directly into a supersonic stream, the pressure recovery in the tube is modified by the presence of a curved shock wave ahead of the tube, which close to the axial streamline may be regarded as plane, Fig. 4.

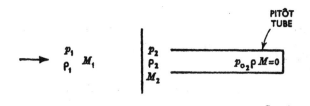

Fig. 4

BOW SHOCK ASSUMED PLANE AND NORMAL TO AXIAL FLOW

The axial flow into the tube may be assumed therefore to have been brought to rest at pressure $p_{0,2}$ from the subsonic regime behind the shock (p_2) *after* it was compressed from the supersonic regime (p_1) by the shock wave.

Applying the isentropic relationships from §2,1 equation (25)

$$\frac{p_{0,2}}{p_2} = \left[1 + \frac{\gamma-1}{2} M_2^2\right]^{\gamma/(\gamma-1)}$$

From the plane normal shock relationships §6,2, equation (77)

$$\frac{p_2}{p_1} = \frac{2\gamma M_1^2 - (\gamma-1)}{\gamma+1}$$

and §6,3, equation (79),

$$M_2^2 = \frac{(\gamma-1)M_1^2 + 2}{2\gamma M_1^2 - (\gamma-1)}$$

Eliminating p_2 and M_2 gives the ratio of the total pressure in a 'supersonic pitot tube' to the static pressure of the (supersonic) free stream :

$$\frac{p_{0,2}}{p_1} = \left[\frac{(\gamma+1)}{2} M_1^2\right]^{\gamma/(\gamma-1)} \bigg/ \left[\frac{2\gamma M_1^2 - (\gamma-1)}{\gamma+1}\right]^{1/(\gamma-1)} \qquad (86)$$

7. PLANE OBLIQUE SHOCK FLOW

Air passing through a plane oblique shock wave is deflected through an angle δ, which is a function of the incident Mach number M_1, the shock wave angle β and the value of γ for the air, Fig. 5.

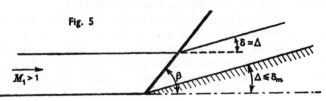

Fig. 5

Full analysis of the relevant equations reveals that for given M_1 and γ:

(a) there is a limiting deflexion angle, δ_m which cannot be exceeded,

(b) a given deflexion angle $\delta < \delta_m$ may be produced by two physically possible shock waves, with different values for the shock angle β. There is a third value of β which, while mathematically possible, is inadmissible on physical grounds, since it corresponds to a *decrease* of entropy.

Now consider a wedge of semi-vertex angle Δ placed with the apex pointing symmetrically into a supersonic stream of given M_1 and γ. The three possibilities, (i) $\Delta < \delta_m$, (ii) $\Delta = \delta_m$, (iii) $\Delta > \delta_m$, are considered in turn.

(i) Experience shows that if $\Delta < \delta_m$ a plane oblique shock wave is formed, attached to the wedge at its apex, Fig. 5. It is found that the shock wave which is formed has the smaller of the two possible values of β. This shock wave is described as 'weak'.

(ii) When $\Delta = \delta_m$ the two physically possible shock waves become identical and it is this shock wave which is formed, attached to the wedge apex. This is a limiting case of (i).

(iii) It has been stated that the flow deflexion angle cannot exceed a certain limiting value δ_m. If the wedge angle Δ is greater than δ_m a plane oblique shock wave cannot be formed. Instead a curved detached shock wave will be formed, as in Fig. 6.

Suppose, for case (iii), the wedge to be the forepart of a body of finite length, such as a blunt or round-nosed aerofoil. On the axis of symmetry the shock wave is normal to the flow and the deflexion angle δ is zero. Away from the axis the flow deflexion angle becomes finite. On the axis, and for a certain distance on either side of the axis, the flow behind the shock wave is subsonic, and the streamlines round the body are those for a subsonic compressible

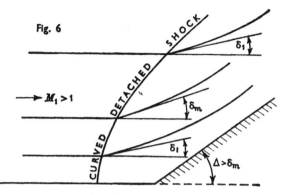

Fig. 6

flow, the initial conditions for which are determined by the flow conditions (Mach number, direction, etc.) immediately behind the shock wave.

Considering points on the shock wave progressively further from the axis, the deflexion angle increases from zero on the axis to δ_m at some point. Over this region a given deflexion is produced by the 'strong shock', i.e. the shock wave with the larger value of β. At the point where $\delta = \delta_m$, the shock wave is the single possible one (as in (ii) above). Moving further from the axis, δ decreases to zero at infinity. Over this outer portion of the shock wave a given value of δ is produced by the 'weak shock', i.e. that corresponding to the smaller of the two possible values of β. Thus, in general, on a curved shock wave there are two points on each side of the axis at which the flow deflexion has a given value, δ_1 say. The point nearer the axis corresponds to the 'strong' shock for the given M_1, δ_1 and γ. At the point further from the axis the shock wave is the 'weak' solution.

From the foregoing it is seen that both solutions for β, given M_1, γ and δ, may be found in practice, as may also the single solution corresponding to M_1, γ and δ_m.

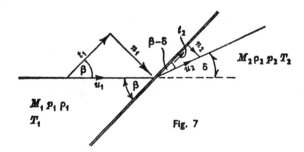

Fig. 7

7,1. The Oblique Shock Equations. Since there is no momentum transfer along the shock front, the plane *oblique* shock wave may be analysed as a plane *normal* shock wave upon which a uniform velocity stream has been superimposed.

In the notation of Fig. 7

$$t_1 = t_2 \quad \cdot \quad \cdot \quad \cdot \quad \cdot \quad \cdot \quad (87)$$

and using the results of the previous section directly with M_1 replaced by $M_1 \sin \beta$ (i.e. the normal component),

$$\frac{p_2}{p_1} = \frac{2\gamma M_1^2 \sin^2\beta - (\gamma - 1)}{\gamma + 1} \quad \cdot \quad \cdot \quad \cdot \quad (88)$$

$$\frac{\rho_2}{\rho_1} = \frac{(\gamma + 1)M_1^2 \sin^2\beta}{2 + (\gamma - 1)M_1^2 \sin^2\beta} \quad \cdot \quad \cdot \quad \cdot \quad (89)$$

$$\frac{n_2}{n_1} = \frac{2 + (\gamma - 1)M_1^2 \sin^2\beta}{(\gamma + 1)M_1^2 \sin^2\beta} \quad \cdot \quad \cdot \quad \cdot \quad (90)$$

Combining equations (88) and (89) gives the temperature ratio

$$\frac{T_2}{T_1} = \frac{2\gamma M_1^2 \sin^2\beta - (\gamma - 1)}{\gamma + 1} \cdot \frac{2 + (\gamma - 1)M_1^2 \sin^2 \beta}{(\gamma + 1)M_1^2 \sin^2 \beta} \quad (91)$$

Likewise from equation (79)

$$M_2^2 = \frac{(\gamma - 1)M_1^2 \sin^2 \beta + 2}{2\gamma M_1^2 \sin^2 \beta - (\gamma - 1)} \cdot \mathrm{cosec}^2 \, (\beta - \delta) \quad \cdot \quad \cdot \quad (92)$$

The entropy follows from substituting the appropriate ratios in equation (11B) §6,6.

The pressure recovery factor $p_{0,2}/p_{0,1}$, is given by

$$R_{0,1} = \frac{p_{0,2}}{p_{0,2}} = \frac{p_{0,2}}{p_2} \cdot \frac{p_2}{p_1} \cdot \frac{p_1}{p_{0,1}} \quad \cdot \quad \cdot \quad \cdot \quad (93)$$

p_2/p_1 is given in Table VI while $p_{0,2}/p_2$ and $p_{0,1}/p_1$ are obtained from Table I, for the known values of M_1 and M_2.

Now by continuity and geometry

$$\frac{\rho_2}{\rho_1} = \frac{\tan \beta}{\tan (\beta - \delta)} \quad \cdot \quad \cdot \quad \cdot \quad \cdot \quad (94)$$

and equating equations (89) and (94) gives on rearranging,

$$\tan \delta = \cot \beta \cdot \frac{M_1^2 \sin^2 \beta - 1}{\dfrac{\gamma + 1}{2} M_1^2 - (M_1^2 \sin^2 \beta - 1)} \quad \cdot \quad (95)$$

Equation (95) connects the deflexion angle δ, incident Mach number M_1 and resultant shock angle inclination β. To solve equation (95) directly to find β, for given M_1 and δ (the case usually met in practice) is very difficult. However Collar has shown† that, writing x for $\cot \beta$, the equation may be rearranged to yield the iterative process

$$x_{n+1} = +\sqrt{A - \frac{B}{x_n + C}} \qquad \ldots \qquad (96)$$

where

$$A = M_1{}^2 - 1$$

$$B = \tfrac{1}{2}(\gamma + 1)M_1{}^4 \tan \delta$$

and

$$C = [\tfrac{1}{2}(\gamma + 1)M_1{}^2 + 1] \tan \delta$$

and that a suitable first approximation x_1 is $\sqrt{M_1{}^2 - 1}$. This method gives, on completion of the iteration,‡ the value of $x = \cot \beta$ corresponding to the weak shock, i.e. the smaller value of β or the larger value of $\cot \beta$.

In the paper quoted, Collar also shows that the equation may be written as the cubic

$$x^3 + Cx^2 - Ax + (B - AC) = 0 \qquad \ldots \qquad (97)$$

The iteration process gives one root, x_0, of the cubic. If this root is then extracted as a factor, the result is the quadratic equation $x^2 + (C + x_0)x + [x_0(C + x_0) - A] = 0$ the formal solution to which is

$$x = \tfrac{1}{2}\left[-(C + x_0) \pm \sqrt{(C + x_0)(C - 3x_0) + 4A}\right] \qquad . \quad . \quad (98)$$

Now it is known that the original cubic has two positive roots and one negative root, and x_0 is one of the positive roots. It therefore follows that the second physically possible solution is that given by the positive root of the quadratic equation (97) and that this root is the value of $\cot \beta$ for the strong shock. Thus it is possible to find the two values of β for given M_1 and δ, provided $\delta < \delta_m$. The limiting value δ_m must be found by a different method.

Direct differentiation of equation (95) with reference to β with M_1 constant gives that, for the maximum value of $\tan \delta$

$$\sin^2 \beta_m = \frac{1}{\gamma M_1{}^2}\left[\frac{\gamma + 1}{4} M_1{}^2 - 1 \right.$$
$$\left. + \sqrt{(\gamma + 1)\left(1 + \frac{\gamma - 1}{2} M_1{}^2 + \frac{\gamma + 1}{16} M_1{}^4\right)}\right] \qquad (99)$$

† Collar, A. R., *J.R.Ae.S.*, November 1959.

‡ The iteration was continued until the difference between two successive values of x was less than 1×10^{-7}.

Using this equation, the value of sin β_m can be determined. Substituting this back into equation (95) then gives the corresponding value for δ, which is the maximum value δ_m.

Having in this way determined all the required values of δ and β for a given incident Mach number, the other quantities (pressure, density, etc.) can be calculated from the appropriate equations.

EXPLANATION OF TABLE PARAMETERS

Tables I. Isentropic flow of dry air with Prandtl-Meyer expansion angles

M Local Mach number, 0·01 (0·01) 4·00

$\left.\begin{array}{l} p_0/p \\ \rho_0/\rho \\ T_0/T \end{array}\right\}$ Ratio of total (or reservoir) to local $\left\{\begin{array}{l} \text{pressure} \\ \text{density} \\ \text{temperature} \end{array}\right.$

A/A^* Ratio of local to throat (sonic) stream-tube area

u/c Ratio of local to ultimate air velocity

$\left.\begin{array}{l} \mu \\ \theta \\ \nu \end{array}\right.$ $\begin{array}{l} \text{Mach angle, degrees} \\ \text{Mach line rotation from datum, degrees} \\ \text{Flow deflexion from datum, degrees} \end{array}\left.\begin{array}{l} \\ \\ \end{array}\right\}\begin{array}{l} \text{expansion} \\ \text{round a} \\ \text{corner} \end{array}$

Table II. Frictionless flow in a constant area duct, with heat transfer (Rayleigh flow)

Note that in this Table the increments of entropy are algebraically negative, due to M = unity being a maximum limiting stage.

M Local Mach number, 0 (0·01) 4·00

$\left(1-\dfrac{q}{c_p T_0^*}\right) = \begin{array}{l} p/p^* \\ T/T^* \\ T_0/T_0^* \end{array}$ $\left.\begin{array}{l} \\ \\ \\ \\ \end{array}\right\}$ Ratio of local to throat (sonic) $\left\{\begin{array}{l} \text{static pressure} \\ \text{static temperature} \\ \text{total temperature} \\ \text{(density) velocity} \\ \text{total pressure} \end{array}\right.$

$(\rho^*/\rho) = u/u^*$

p_0/p_0^*

ΔS Entropy change (from conditions at M = unity)

Note. $\dfrac{q}{c_p T_0^*}$ is a non-dimensional quantity.

Table III. Adiabatic flow in a constant area duct with surface friction (Fanno flow)

Again, a limiting state occurs at M = unity, and local increments of entropy are algebraically negative.

M Local Mach number, 0·01 (0·01) 4·00

$(u^*/u) = \begin{array}{l} p/p^* \\ p_0/p_0^* \\ \rho/\rho^* \\ T/T^* \end{array}$ $\left.\begin{array}{l} \\ \\ \\ \end{array}\right\}$ Ratio of local to throat (sonic) $\left\{\begin{array}{l} \text{static pressure} \\ \text{total pressure} \\ \text{(velocity) density} \\ \text{static temperature} \end{array}\right.$

ΔS Entropy change (from conditions at M = unity)

$\dfrac{C_f 4|x-x^*|}{d}$ Non-dimensional shear force quantity, i.e. four times the product of the friction coefficient and that pipe length (in pipe diameters) required to produce sonic conditions.

Table IV. Plane normal shock wave

Suffices 1 and 2 denote conditions respectively before and behind the plane normal shock wave. An additional suffix 0 denotes total values, i.e. values attained by bringing the air to rest isentropically. The quantity $p_{0,2}/p_1$ is often known as 'Rayleigh's supersonic pitôt-tube relationship'. The entropy values given are increases from the value for the undisturbed stream.

M_1	Local initial Mach number, $1\cdot01$ $(0\cdot01)$ $4\cdot00$
p_2/p_1	Ratio of static pressures across shock wave
ρ_2/ρ_1	Ratio of densities across shock wave
T_2/T_1	Ratio of static temperatures across shock wave
M_2	Exit Mach number
u_2/u_1	Ratio of velocities across shock wave
ΔS	Entropy increase across shock wave
$p_{0,2}/p_1$	Ratio of total pressure behind, to static pressure ahead

Table V. Plane oblique shock wave

The flow deflexion angle δ and the initial Mach number M_1 have been taken as the independent variables in this Table. At each value of M_1, δ increases from zero (these values are in fact given in Table IV) in 2° steps up to the maximum flow deflexion for that initial Mach number. The results appearing before the maximum deflexion are for the 'weak' shock and those after the maximum deflexion are for the 'strong' shock.

Suffices 1 and 2 denote conditions respectively before and behind the shock wave; the additional suffix 0 denotes total conditions.

M_1	Initial Mach number, $1\cdot05$ $(0\cdot05)$ $3\cdot95$
δ	Flow deflexion angle, degrees
β	Shock wave angle to free stream, degrees
p_2/p_1	Ratio of static pressures across shock wave
ρ_2/ρ_1	Ratio of densities across shock wave
T_2/T_1	Ratio of static temperatures across shock wave
M_2	Exit Mach number
ΔS	Entropy increase across shock wave

TABLE I A

ISENTROPIC FLOW OF DRY AIR WITH PRANDTL–MEYER EXPANSION ANGLES
(FOR EQUAL INCREMENTS OF MACH NUMBER FROM 0 TO 4.00)

M	p_0/p	ρ_0/ρ	T_0/T	A/A^*	u/c	μ	θ	v
0·01	1·000	1·000	1·000	57·855	0·0045	These headings are		
0·02	1·000	1·000	1·000	28·933	0·0090	not applicable for		
0·03	1·001	1·000	1·000	19·294	0·0135	Mach numbers less		
0·04	1·001	1·001	1·000	14·477	0·0180	than unity		
0·05	1·002	1·001	1·001	11·588	0·0224			
0·06	1·003	1·002	1·001	9·663	0·0269			
0·07	1·003	1·002	1·001	8·289	0·0314			
0·08	1·004	1·003	1·001	7·259	0·0359			
0·09	1·006	1·004	1·002	6·459	0·0404			
0·10	1·007	1·005	1·002	5·820	0·0448			
0·11	1·009	1·006	1·002	5·298	0·0493			
0·12	1·010	1·007	1·003	4·863	0·0538			
0·13	1·012	1·008	1·003	4·495	0·0583			
0·14	1·014	1·010	1·004	4·181	0·0627			
0·15	1·016	1·011	1·005	3·909	0·0672			
0·16	1·018	1·013	1·005	3·672	0·0716			
0·17	1·020	1·015	1·006	3·462	0·0761			
0·18	1·023	1·016	1·007	3·277	0·0805			
0·19	1·026	1·018	1·007	3·111	0·0850			
0·20	1·028	1·020	1·008	2·963	0·0894			
0·21	1·031	1·022	1·009	2·828	0·0939			
0·22	1·034	1·024	1·010	2·707	0·0983			
0·23	1·038	1·027	1·011	2·596	0·1027			
0·24	1·041	1·029	1·012	2·495	0·1071			
0·25	1·045	1·032	1·013	2·402	0·1115			
0·26	1·048	1·034	1·014	2·317	0·1159			
0·27	1·052	1·037	1·015	2·238	0·1203			
0·28	1·056	1·040	1·016	2·165	0·1247			
0·29	1·060	1·043	1·017	2·097	0·1291			
0·30	1·065	1·046	1·018	2·035	0·1335			
0·31	1·069	1·049	1·019	1·976	0·1378			
0·32	1·074	1·052	1·021	1·921	0·1422			
0·33	1·078	1·055	1·022	1·870	0·1465			
0·34	1·083	1·059	1·023	1·822	0·1509			
0·35	1·089	1·062	1·025	1·778	0·1552			
0·36	1·094	1·066	1·026	1·735	0·1595			
0·37	1·099	1·070	1·028	1·696	0·1638			
0·38	1·105	1·074	1·029	1·658	0·1681			
0·39	1·111	1·078	1·031	1·623	0·1724			
0·40	1·117	1·082	1·032	1·590	0·1767			
0·41	1·123	1·086	1·034	1·558	0·1810			
0·42	1·129	1·091	1·036	1·529	0·1853			
0·43	1·136	1·095	1·037	1·500	0·1895			
0·44	1·143	1·100	1·039	1·474	0·1938			
0·45	1·149	1·104	1·041	1·448	0·1980			

TABLE I A

ISENTROPIC FLOW OF DRY AIR WITH PRANDTL—MEYER EXPANSION ANGLES
(FOR EQUAL INCREMENTS OF MACH NUMBER FROM 0 TO 4.00)

M	p_0/p	ρ_0/ρ	T_0/T	A/A^*	u/c	μ	θ	v
0·46	1·156	1·109	1·043	1·424	0·2022	These headings are		
0·47	1·164	1·114	1·045	1·402	0·2064	not applicable for		
0·48	1·171	1·119	1·046	1·380	0·2106	Mach numbers less		
0·49	1·179	1·124	1·048	1·359	0·2148	than unity		
0·50	1·187	1·130	1·050	1·340	0·2190			
0·51	1·195	1·135	1·052	1·321	0·2232			
0·52	1·203	1·141	1·054	1·303	0·2273			
0·53	1·211	1·146	1·057	1·286	0·2315			
0·54	1·220	1·152	1·059	1·270	0·2356			
0·55	1·229	1·158	1·061	1·255	0·2397			
0·56	1·238	1·164	1·063	1·240	0·2438			
0·57	1·247	1·170	1·065	1·226	0·2479			
0·58	1·257	1·177	1·068	1·213	0·2520			
0·59	1·266	1·183	1·070	1·200	0·2560			
0·60	1·276	1·190	1·073	1·188	0·2601			
0·61	1·286	1·197	1·075	1·177	0·2641			
0·62	1·297	1·203	1·077	1·166	0·2681			
0·63	1·307	1·210	1·080	1·155	0·2721			
0·64	1·318	1·217	1·083	1·145	0·2761			
0·65	1·329	1·225	1·085	1·136	0·2801			
0·66	1·340	1·232	1·088	1·126	0·2841			
0·67	1·352	1·240	1·090	1·118	0·2880			
0·68	1·364	1·247	1·093	1·110	0·2919			
0·69	1·376	1·255	1·096	1·102	0·2959			
0·70	1·388	1·263	1·099	1·094	0·2998			
0·71	1·400	1·271	1·102	1·087	0·3037			
0·72	1·413	1·280	1·104	1·080	0·3075			
0·73	1·426	1·288	1·107	1·074	0·3114			
0·74	1·440	1·297	1·110	1·068	0·3152			
0·75	1·453	1·305	1·113	1·062	0·3191			
0·76	1·467	1·314	1·116	1·057	0·3229			
0·77	1·481	1·323	1·119	1·052	0·3267			
0·78	1·496	1·332	1·123	1·047	0·3305			
0·79	1·510	1·342	1·126	1·042	0·3342			
0·80	1·525	1·351	1·129	1·038	0·3380			
0·81	1·541	1·361	1·132	1·034	0·3417			
0·82	1·556	1·371	1·135	1·030	0·3454			
0·83	1·572	1·381	1·139	1·027	0·3491			
0·84	1·589	1·391	1·142	1·024	0·3528			
0·85	1·605	1·401	1·146	1·021	0·3565			
0·86	1·622	1·412	1·149	1·018	0·3601			
0·87	1·639	1·422	1·153	1·015	0·3638			
0·88	1·657	1·433	1·156	1·013	0·3674			
0·89	1·675	1·444	1·160	1·011	0·3710			
0·90	1·693	1·455	1·163	1·009	0·3746			

TABLE I A

ISENTROPIC FLOW OF DRY AIR WITH PRANDTL-MEYER EXPANSION ANGLES
(FOR EQUAL INCREMENTS OF MACH NUMBER FROM 0 TO 4.00)

M	p_0/p	ρ_0/ρ	T_0/T	A/A^*	u/c	μ	θ	v
0·91	1·711	1·467	1·167	1·007	0·3782	These headings are		
0.92	1·730	1·478	1·171	1·006	0·3817	not applicable for		
0·93	1·749	1·490	1·174	1·004	0·3852	Mach numbers less		
0·94	1·769	1·502	1·178	1·003	0·3888	than unity		
0·95	1·789	1·514	1·182	1·002	0·3923			
0·96	1·809	1·526	1·186	1·001	0·3957			
0·97	1·830	1·538	1·190	1·001	0·3992			
0·98	1·851	1·551	1·194	1·000	0·4027			
0·99	1·873	1·564	1·197	1·000	0·4061			
1·00	1·895	1·577	1·201	1·000	0·4095	90·00	0·00	0·00
1·01	1·917	1·590	1·206	1·000	0·4129	81·93	8·11	0·04
1·02	1·940	1·604	1·210	1·000	0·4163	78·64	11·49	0·13
1·03	1·963	1·617	1·214	1·001	0·4197	76·14	14·09	0·23
1·04	1·987	1·631	1·218	1·001	0·4230	74·06	16·29	0·35
1·05	2·011	1·645	1·222	1·002	0·4263	72·25	18·24	0·49
1·06	2·035	1·659	1·226	1·003	0·4297	70·63	20·01	0·64
1·07	2·060	1·674	1·231	1·004	0·4330	69·16	21·64	0·80
1·08	2·085	1·688	1·235	1·005	0·4362	67·81	23·16	0·97
1·09	2·111	1·703	1·239	1·006	0·4395	66·55	24·59	1·15
1·10	2·137	1·718	1·244	1·008	0·4427	65·38	25·95	1·33
1·11	2·164	1·734	1·248	1·010	0·4460	64·28	27·25	1·53
1·12	2·191	1·749	1·253	1·011	0·4492	63·23	28·50	1·73
1·13	2·219	1·765	1·257	1·013	0·4524	62·25	29·70	1·94
1·14	2·247	1·781	1·262	1·015	0·4555	61·31	30·85	2·16
1·15	2·276	1·797	1·266	1·017	0·4587	60·41	31·97	2·38
1·16	2·305	1·814	1·271	1·020	0·4618	59·55	33·05	2·60
1·17	2·335	1·830	1·276	1·022	0·4650	58·73	34·11	2·83
1·18	2·365	1·847	1·281	1·025	0·4681	57·94	35·13	3·07
1·19	2·396	1·864	1·285	1·028	0·4712	57·18	36·13	3·31
1·20	2·428	1·882	1·290	1·030	0·4742	56·44	37·11	3·55
1·21	2·460	1·899	1·295	1·033	0·4773	55·74	38·06	3·80
1·22	2·492	1·917	1·300	1·037	0·4803	55·05	39·00	4·05
1·23	2·525	1·935	1·305	1·040	0·4834	54·39	39·91	4·31
1·24	2·559	1·954	1·310	1·043	0·4864	53·75	40·81	4·56
1·25	2·593	1·972	1·315	1·047	0·4893	53·13	41·69	4·82
1·26	2·628	1·991	1·320	1·050	0·4923	52·53	42·56	5·09
1·27	2·664	2·010	1·325	1·054	0·4953	51·94	43·41	5·35
1·28	2·700	2·030	1·330	1·058	0·4982	51·38	44·24	5·62
1·29	2·737	2·049	1·335	1·062	0·5011	50·82	45·07	5·89
1·30	2·774	2·069	1·341	1·066	0·5040	50·28	45·88	6·16
1·31	2·812	2·090	1·346	1·070	0·5069	49·76	46·67	6·43
1·32	2·851	2·110	1·351	1·075	0·5098	49·25	47·46	6·71
1·33	2·890	2·131	1·356	1·079	0·5126	48·75	48·23	6·99
1·34	2·930	2·152	1·362	1·084	0·5154	48·27	49·00	7·27
1·35	2·971	2·173	1·367	1·089	0·5183	47·79	49·75	7·55

TABLE I A

ISENTROPIC FLOW OF DRY AIR WITH PRANDTL—MEYER EXPANSION ANGLES
(FOR EQUAL INCREMENTS OF MACH NUMBER FROM 0 TO 4.00)

M	p_0/p	ρ_0/ρ	T_0/T	A/A^*	u/c	μ	θ	ν
1·36	3·013	2·195	1·373	1·094	0·5211	47·33	50·50	7·83
1·37	3·055	2·217	1·378	1·099	0·5238	46·88	51·23	8·11
1·38	3·098	2·239	1·384	1·104	0·5266	46·44	51·96	8·40
1·39	3·142	2·261	1·389	1·109	0·5294	46·01	52·68	8·68
1·40	3·186	2·284	1·395	1·115	0·5321	45·58	53·39	8·97
1·41	3·231	2·307	1·401	1·120	0·5348	45·17	54·09	9·26
1·42	3·277	2·330	1·406	1·126	0·5375	44·77	54·78	9·55
1·43	3·324	2·354	1·412	1·132	0·5402	44·37	55·47	9·84
1·44	3·372	2·378	1·418	1·138	0·5429	43·98	56·15	10·13
1·45	3·420	2·402	1·424	1·144	0·5455	43·60	56·82	10·42
1·46	3·470	2·427	1·430	1·150	0·5481	43·23	57·48	10·71
1·47	3·520	2·452	1·435	1·156	0·5508	42·86	58·14	11·00
1·48	3·571	2·477	1·441	1·163	0·5534	42·51	58·79	11·30
1·49	3·623	2·503	1·447	1·169	0·5560	42·16	59·43	11·59
1·50	3·675	2·529	1·453	1·176	0·5585	41·81	60·07	11·88
1·51	3·729	2·555	1·459	1·183	0·5611	41·47	60·71	12·18
1·52	3·784	2·582	1·466	1·190	0·5636	41·14	61·33	12·47
1·53	3·839	2·609	1·472	1·197	0·5661	40·81	61·95	12·77
1·54	3·896	2·636	1·478	1·204	0·5686	40·49	62·57	13·06
1·55	3·953	2·664	1·484	1·211	0·5711	40·18	63·18	13·36
1·56	4·011	2·692	1·490	1·219	0·5736	39·87	63·78	13·65
1·57	4·071	2·720	1·497	1·226	0·5761	39·56	64·38	13·95
1·58	4·131	2·749	1·503	1·234	0·5785	39·27	64·98	14·24
1·59	4·193	2·778	1·509	1·242	0·5809	38·97	65·56	14·54
1·60	4·255	2·807	1·516	1·250	0·5834	38·68	66·15	14·83
1·61	4·319	2·837	1·522	1·258	0·5857	38·40	66·73	15·13
1·62	4·383	2·867	1·529	1·266	0·5881	38·12	67·30	15·42
1·63	4·449	2·898	1·535	1·274	0·5905	37·84	67·87	15·72
1·64	4·516	2·929	1·542	1·283	0·5929	37·57	68·44	16·01
1·65	4·584	2·960	1·549	1·292	0·5952	37·31	69·00	16·30
1·66	4·653	2·992	1·555	1·300	0·5975	37·04	69·56	16·60
1·67	4·723	3·024	1·562	1·309	0·5998	36·78	70·11	16·89
1·68	4·795	3·056	1·569	1·318	0·6021	36·53	70·66	17·19
1·69	4·867	3·089	1·576	1·327	0·6044	36·28	71·20	17·48
1·70	4·941	3·123	1·582	1·337	0·6066	36·03	71·74	17·77
1·71	5·016	3·157	1·589	1·346	0·6089	35·79	72·28	18·07
1·72	5·093	3·191	1·596	1·356	0·6111	35·55	72·81	18·36
1·73	5·170	3·225	1·603	1·366	0·6133	35·31	73·34	18·65
1·74	5·249	3·260	1·610	1·375	0·6156	35·08	73·86	18·94
1·75	5·330	3·296	1·617	1·385	0·6177	34·85	74·38	19·23
1·76	5·411	3·332	1·624	1·396	0·6199	34·62	74·90	19·52
1·77	5·494	3·368	1·631	1·406	0·6221	34·40	75·41	19·81
1·78	5·578	3·405	1·638	1·416	0·6242	34·18	75·92	20·10
1·79	5·664	3·442	1·646	1·427	0·6264	33·96	76·43	20·39
1·80	5·751	3·480	1·653	1·438	0·6285	33·75	76·93	20·68

4

TABLE I A

ISENTROPIC FLOW OF DRY AIR WITH PRANDTL-MEYER EXPANSION ANGLES
(FOR EQUAL INCREMENTS OF MACH NUMBER FROM 0 TO 4.00)

M	p_0/p	ρ_0/ρ	T_0/T	A/A^*	u/c	μ	θ	v
1·81	5·840	3·518	1·660	1·449	0·6306	33·54	77·43	20·97
1·82	5·930	3·556	1·667	1·460	0·6327	33·33	77·93	21·26
1·83	6·022	3·595	1·675	1·471	0·6348	33·12	78·42	21·54
1·84	6·115	3·635	1·682	1·482	0·6368	32·92	78·91	21·83
1·85	6·209	3·675	1·690	1·494	0·6389	32·72	79·39	22·11
1·86	6·305	3·715	1·697	1·505	0·6409	32·52	79·88	22·40
1·87	6·403	3·756	1·705	1·517	0·6429	32·33	80·36	22·68
1·88	6·502	3·798	1·712	1·529	0·6449	32·13	80·83	22·97
1·89	6·603	3·840	1·720	1·541	0·6469	31·94	81·31	23·25
1·90	6·706	3·882	1·727	1·554	0·6489	31·76	81·78	23·53
1·91	6·810	3·925	1·735	1·566	0·6509	31·57	82·24	23·81
1·92	6·916	3·969	1·743	1·579	0·6529	31·39	82·71	24·09
1·93	7·024	4·013	1·751	1·591	0·6548	31·21	83·17	24·38
1·94	7·134	4·057	1·758	1·604	0·6567	31·03	83·63	24·65
1·95	7·245	4·102	1·766	1·617	0·6586	30·85	84·08	24·93
1·96	7·358	4·148	1·774	1·631	0·6606	30·68	84·53	25·21
1·97	7·473	4·194	1·782	1·644	0·6624	30·51	84·98	25·49
1·98	7·590	4·240	1·790	1·658	0·6643	30·33	85·43	25·77
1·99	7·709	4·288	1·798	1·671	0·6662	30·17	85·88	26·04
2·00	7·830	4·335	1·806	1·685	0·6680	30·00	86·32	26·32
2·01	7·952	4·384	1·814	1·699	0·6699	29·84	86·76	26·59
2·02	8·077	4·432	1·822	1·714	0·6717	29·67	87·19	26·86
2·03	8·203	4·482	1·830	1·728	0·6735	29·51	87·63	27·14
2·04	8·332	4·532	1·839	1·743	0·6753	29·35	88·06	27·41
2·05	8·463	4·582	1·847	1·757	0·6771	29·20	88·48	27·68
2·06	8·596	4·634	1·855	1·772	0·6789	29·04	88·91	27·95
2·07	8·731	4·685	1·863	1·787	0·6807	28·89	89·33	28·22
2·08	8·868	4·738	1·872	1·803	0·6825	28·74	89·75	28·49
2·09	9·007	4·791	1·880	1·818	0·6842	28·59	90·17	28·76
2·10	9·149	4·844	1·889	1·834	0·6859	28·44	90·59	29·02
2·11	9·293	4·898	1·897	1·850	0·6877	28·29	91·00	29·29
2·12	9·439	4·953	1·906	1·866	0·6894	28·14	91·41	29·56
2·13	9·587	5·009	1·914	1·882	0·6911	28·00	91·82	29·82
2·14	9·738	5·065	1·923	1·898	0·6928	27·86	92·23	30·09
2·15	9·891	5·121	1·931	1·915	0·6944	27·72	92·63	30·35
2·16	10·047	5·179	1·940	1·932	0·6961	27·58	93·03	30·61
2·17	10·205	5·237	1·949	1·949	0·6978	27·44	93·43	30·87
2·18	10·366	5·295	1·958	1·966	0·6994	27·30	93·83	31·13
2·19	10·529	5·355	1·966	1·983	0·7010	27·17	94·22	31·39
2·20	10·695	5·415	1·975	2·001	0·7027	27·04	94·62	31·65
2·21	10·864	5·475	1·984	2·019	0·7043	26·90	95·01	31·91
2·22	11·035	5·536	1·993	2·037	0·7059	26·77	95·39	32·17
2·23	11·208	5·598	2·002	2·055	0·7075	26·64	95·78	32·42
2·24	11·385	5·661	2·011	2·073	0·7090	26·51	96·16	32·68
2·25	11·564	5·725	2·020	2·092	0·7106	26·39	96·54	32·93

TABLE I A

ISENTROPIC FLOW OF DRY AIR WITH PRANDTL—MEYER EXPANSION ANGLES
(FOR EQUAL INCREMENTS OF MACH NUMBER FROM 0 TO 4.00)

M	p_0/p	ρ_0/ρ	T_0/T	A/A^*	u/c	μ	θ	v
2·26	11·746	5·789	2·029	2·111	0·7122	26·26	96·92	33·19
2·27	11·931	5·854	2·038	2·130	0·7137	26·14	97·30	33·44
2·28	12·119	5·919	2·047	2·149	0·7153	26·01	97·68	33·69
2·29	12·310	5·985	2·057	2·168	0·7168	25·89	98·05	33·94
2·30	12·504	6·052	2·066	2·188	0·7183	25·77	98·42	34·19
2·31	12·701	6·120	2·075	2·208	0·7198	25·65	98·79	34·44
2·32	12·901	6·189	2·085	2·228	0·7213	25·53	99·16	34·69
2·33	13·104	6·258	2·094	2·248	0·7228	25·42	99·52	34·94
2·34	13·310	6·328	2·103	2·269	0·7243	25·30	99·88	35·18
2·35	13·519	6·399	2·113	2·290	0·7257	25·18	100·25	35·43
2·36	13·732	6·470	2·122	2·311	0·7272	25·07	100·60	35·67
2·37	13·948	6·543	2·132	2·332	0·7286	24·96	100·96	35·92
2·38	14·167	6·616	2·141	2·353	0·7301	24·85	101·32	36·16
2·39	14·389	6·690	2·151	2·375	0·7315	24·73	101·67	36·40
2·40	14·615	6·764	2·161	2·397	0·7329	24·62	102·02	36·65
2·41	14·845	6·840	2·170	2·419	0·7343	24·52	102·37	36·89
2·42	15·078	6·916	2·180	2·441	0·7357	24·41	102·72	37·13
2·43	15·315	6·993	2·190	2·464	0·7371	24·30	103·06	37·36
2·44	15·555	7·072	2·200	2·487	0·7385	24·19	103·41	37·60
2·45	15·799	7·150	2·210	2·510	0·7399	24·09	103·75	37·84
2·46	16·046	7·230	2·219	2·533	0·7412	23·99	104·09	38·08
2·47	16·298	7·311	2·229	2·557	0·7426	23·88	104·43	38·31
2·48	16·553	7·392	2·239	2·580	0·7439	23·78	104·77	38·55
2·49	16·812	7·474	2·249	2·604	0·7453	23·68	105·10	38·78
2·50	17·075	7·558	2·259	2·629	0·7466	23·58	105·43	39·01
2·51	17·343	7·642	2·269	2·653	0·7479	23·48	105·77	39·24
2·52	17·614	7·727	2·280	2·678	0·7492	23·38	106·10	39·48
2·53	17·889	7·812	2·290	2·703	0·75C5	23·28	106·42	39·71
2·54	18·168	7·899	2·300	2·729	0·7518	23·18	106·75	39·94
2·55	18·452	7·987	2·310	2·754	0·7531	23·09	107·08	40·16
2·56	18·740	8·076	2·321	2·780	0·7544	22·99	107·40	40·39
2·57	19·032	8·165	2·331	2·806	0·7556	22·90	107·72	40·62
2·58	19·329	8·256	2·341	2·832	0·7569	22·81	108·04	40·84
2·59	19·630	8·347	2·352	2·859	0·7581	22·71	108·36	41·07
2·60	19·935	8·440	2·362	2·886	0·7594	22·62	108·67	41·29
2·61	20·245	8·533	2·373	2·913	0·7606	22·53	108·99	41·52
2·62	20·560	8·627	2·383	2·941	0·7618	22·44	109·30	41·74
2·63	20·880	8·723	2·394	2·969	0·7631	22·35	109·61	41·96
2·64	21·204	8·819	2·404	2·997	0·7643	22·26	109·92	42·18
2·65	21·533	8·916	2·415	3·025	0·7655	22·17	110·23	42·40
2·66	21·867	9·015	2·426	3·053	0·7667	22·08	110·54	42·62
2·67	22·206	9·114	2·436	3·082	0·7678	22·0O	110·85	42·84
2·68	22·550	9·214	2·447	3·112	0·7690	21·91	111·15	43·06
2·69	22·899	9·316	2·458	3·141	0·7702	21·82	111·45	43·28
2·70	23·253	9·418	2·469	3·171	0·7713	21·74	111·75	43·49

TABLE I A

ISENTROPIC FLOW OF DRY AIR WITH PRANDTL-MEYER EXPANSION ANGLES
(FOR EQUAL INCREMENTS OF MACH NUMBER FROM 0 TO 4.00)

M	p_0/p	ρ_0/ρ	T_0/T	A/A^*	u/c	μ	θ	ν
2.71	23.612	9.522	2.480	3.201	0.7725	21.65	112.05	43.71
2.72	23.977	9.626	2.491	3.231	0.7736	21.57	112.35	43.92
2.73	24.347	9.732	2.502	3.262	0.7748	21.49	112.65	44.13
2.74	24.723	9.839	2.513	3.293	0.7759	21.41	112.94	44.35
2.75	25.104	9.947	2.524	3.324	0.7770	21.32	113.24	44.56
2.76	25.490	10.056	2.535	3.356	0.7781	21.24	113.53	44.77
2.77	25.882	10.166	2.546	3.388	0.7793	21.16	113.82	44.98
2.78	26.280	10.277	2.557	3.420	0.7804	21.08	114.11	45.19
2.79	26.684	10.389	2.568	3.452	0.7815	21.00	114.40	45.40
2.80	27.094	10.502	2.580	3.485	0.7825	20.92	114.68	45.61
2.81	27.509	10.617	2.591	3.518	0.7836	20.85	114.97	45.81
2.82	27.931	10.733	2.602	3.552	0.7847	20.77	115.25	46.02
2.83	28.359	10.850	2.614	3.586	0.7858	20.69	115.53	46.23
2.84	28.792	10.968	2.625	3.620	0.7868	20.62	115.81	46.43
2.85	29.233	11.087	2.637	3.654	0.7879	20.54	116.09	46.63
2.86	29.679	11.207	2.648	3.689	0.7889	20.47	116.37	46.84
2.87	30.132	11.329	2.660	3.724	0.7900	20.39	116.65	47.04
2.88	30.591	11.452	2.671	3.760	0.7910	20.32	116.92	47.24
2.89	31.057	11.576	2.683	3.796	0.7920	20.24	117.20	47.44
2.90	31.530	11.701	2.695	3.832	0.7930	20.17	117.47	47.64
2.91	32.010	11.828	2.706	3.868	0.7940	20.10	117.74	47.84
2.92	32.496	11.956	2.718	3.905	0.7950	20.03	118.01	48.04
2.93	32.989	12.085	2.730	3.942	0.7960	19.96	118.28	48.24
2.94	33.490	12.215	2.742	3.980	0.7970	19.89	118.55	48.43
2.95	33.997	12.347	2.754	4.018	0.7980	19.81	118.81	48.63
2.96	34.512	12.480	2.765	4.056	0.7990	19.75	119.08	48.83
2.97	35.034	12.614	2.777	4.095	0.8000	19.68	119.34	49.02
2.98	35.563	12.749	2.789	4.134	0.8009	19.61	119.61	49.21
2.99	36.100	12.886	2.801	4.173	0.8019	19.54	119.87	49.41
3.00	36.644	13.024	2.813	4.213	0.8029	19.47	120.13	49.60
3.01	37.196	13.164	2.826	4.253	0.8038	19.40	120.39	49.79
3.02	37.756	13.305	2.838	4.293	0.8047	19.34	120.64	49.98
3.03	38.324	13.447	2.850	4.334	0.8057	19.27	120.90	50.17
3.04	38.899	13.591	2.862	4.376	0.8066	19.20	121.16	50.36
3.05	39.483	13.736	2.874	4.417	0.8075	19.14	121.41	50.55
3.06	40.075	13.882	2.887	4.459	0.8085	19.07	121.66	50.74
3.07	40.675	14.030	2.899	4.502	0.8094	19.01	121.91	50.92
3.08	41.283	14.179	2.912	4.544	0.8103	18.95	122.17	51.11
3.09	41.900	14.330	2.924	4.588	0.8112	18.88	122.41	51.30
3.10	42.526	14.482	2.936	4.631	0.8121	18.82	122.66	51.48
3.11	43.160	14.636	2.949	4.675	0.8130	18.76	122.91	51.67
3.12	43.803	14.791	2.961	4.720	0.8138	18.69	123.16	51.85
3.13	44.455	14.948	2.974	4.764	0.8147	18.63	123.40	52.03
3.14	45.116	15.106	2.987	4.810	0.8156	18.57	123.64	52.22
3.15	45.786	15.265	2.999	4.855	0.8165	18.51	123.89	52.40

7

TABLE I A

ISENTROPIC FLOW OF DRY AIR WITH PRANDTL—MEYER EXPANSION ANGLES
(FOR EQUAL INCREMENTS OF MACH NUMBER FROM 0 TO 4.00)

M	p_0/p	ρ_0/ρ	T_0/T	A/A^*	u/c	μ	θ	ν
3·16	46·465	15·426	3·012	4·901	0·8173	18·45	124·13	52·5
3·17	47·154	15·589	3·025	4·948	0·8182	18·39	124·37	52·7
3·18	47·852	15·753	3·038	4·995	0·8190	18·33	124·61	52·9
3·19	48·560	15·919	3·050	5·042	0·8199	18·27	124·85	53·1
3·20	49·277	16·086	3·063	5·090	0·8207	18·21	125·08	53·2
3·21	50·004	16·255	3·076	5·138	0·8215	18·15	125·32	53·4
3·22	50·741	16·425	3·089	5·187	0·8224	18·09	125·55	53·6
3·23	51·488	16·597	3·102	5·236	0·8232	18·03	125·79	53·8
3·24	52·246	16·771	3·115	5·285	0·8240	17·98	126·02	54·0
3·25	53·013	16·946	3·128	5·335	0·8248	17·92	126·25	54·1
3·26	53·791	17·123	3·141	5·386	0·8256	17·86	126·48	54·3
3·27	54·579	17·301	3·155	5·437	0·8264	17·81	126·71	54·5
3·28	55·379	17·482	3·168	5·488	0·8272	17·75	126·94	54·6
3·29	56·188	17·663	3·181	5·540	0·8280	17·70	127·17	54·8
3·30	57·009	17·847	3·194	5·592	0·8288	17·64	127·40	55·0
3·31	57·841	18·032	3·208	5·645	0·8296	17·58	127·62	55·2
3·32	58·684	18·219	3·221	5·698	0·8304	17·53	127·85	55·3
3·33	59·538	18·408	3·234	5·751	0·8312	17·48	128·07	55·5
3·34	60·404	18·598	3·248	5·805	0·8319	17·42	128·29	55·7
3·35	61·281	18·790	3·261	5·860	0·8327	17·37	128·52	55·8
3·36	62·170	18·984	3·275	5·915	0·8335	17·31	128·74	56·0
3·37	63·071	19·180	3·288	5·971	0·8342	17·26	128·96	56·2
3·38	63·984	19·377	3·302	6·027	0·8350	17·21	129·17	56·3
3·39	64·909	19·577	3·316	6·083	0·8357	17·16	129·39	56·5
3·40	65·846	19·778	3·329	6·140	0·8364	17·10	129·61	56·7
3·41	66·796	19·981	3·343	6·198	0·8372	17·05	129·82	56·8
3·42	67·758	20·185	3·357	6·256	0·8379	17·00	130·04	57·0
3·43	68·733	20·392	3·371	6·314	0·8386	16·95	130·25	57·2
3·44	69·721	20·600	3·384	6·373	0·8394	16·90	130·47	57·3
3·45	70·722	20·811	3·398	6·433	0·8401	16·85	130·68	57·5
3·46	71·736	21·023	3·412	6·493	0·8408	16·80	130·89	57·6
3·47	72·763	21·237	3·426	6·554	0·8415	16·75	131·10	57·8
3·48	73·803	21·453	3·440	6·615	0·8422	16·70	131·31	58·0
3·49	74·858	21·671	3·454	6·676	0·8429	16·65	131·52	58·1
3·50	75·926	21·891	3·468	6·739	0·8436	16·60	131·73	58·3
3·51	77·008	22·113	3·483	6·801	0·8443	16·55	131·93	58·4
3·52	78·104	22·337	3·497	6·865	0·8450	16·50	132·14	58·6
3·53	79·214	22·562	3·511	6·928	0·8457	16·46	132·34	58·8
3·54	80·339	22·790	3·525	6·993	0·8464	16·41	132·55	58·9
3·55	81·478	23·020	3·539	7·058	0·8470	16·36	132·75	59·1
3·56	82·632	23·252	3·554	7·123	0·8477	16·31	132·95	59·2
3·57	83·801	23·486	3·568	7·189	0·8484	16·27	133·15	59·4
3·58	84·985	23·722	3·583	7·256	0·8490	16·22	133·35	59·5
3·59	86·184	23·960	3·597	7·323	0·8497	16·17	133·55	59·7
3·60	87·398	24·200	3·611	7·390	0·8504	16·13	133·75	59·8

TABLE I A

ISENTROPIC FLOW OF DRY AIR WITH PRANDTL-MEYER EXPANSION ANGLES
(FOR EQUAL INCREMENTS OF MACH NUMBER FROM 0 TO 4.00)

M	p_0/p	ρ_0/ρ	T_0/T	A/A	u/c	μ	θ	v
3·61	88·628	24·443	3·626	7·459	0·8510	16·08	133·95	60·03
3·62	89·874	24·687	3·641	7·528	0·8517	16·04	134·15	60·18
3·63	91·136	24·934	3·655	7·597	0·8523	15·99	134·34	60·34
3·64	92·414	25·182	3·670	7·667	0·8529	15·95	134·54	60·49
3·65	93·708	25·433	3·684	7·738	0·8536	15·90	134·73	60·64
3·66	95·019	25·686	3·699	7·809	0·8542	15·86	134·93	60·78
3·67	96·346	25·941	3·714	7·881	0·8548	15·81	135·12	60·93
3·68	97·690	26·199	3·729	7·953	0·8555	15·77	135·31	61·08
3·69	99·052	26·459	3·744	8·026	0·8561	15·72	135·51	61·23
3·70	100·430	26·721	3·759	8·100	0·8567	15·68	135·70	61·38
3·71	101·826	26·985	3·773	8·174	0·8573	15·64	135·89	61·52
3·72	103·239	27·251	3·788	8·249	0·8579	15·59	136·07	61·67
3·73	104·671	27·520	3·803	8·324	0·8585	15·55	136·26	61·81
3·74	106·120	27·791	3·819	8·400	0·8591	15·51	136·45	61·96
3·75	107·587	28·064	3·834	8·477	0·8597	15·47	136·64	62·10
3·76	109·073	28·340	3·849	8·554	0·8603	15·42	136·82	62·25
3·77	110·577	28·618	3·864	8·632	0·8609	15·38	137·01	62·39
3·78	112·100	28·898	3·879	8·711	0·8615	15·34	137·19	62·53
3·79	113·642	29·181	3·894	8·790	0·8621	15·30	137·38	62·68
3·80	115·204	29·466	3·910	8·870	0·8627	15·26	137·56	62·82
3·81	116·784	29·754	3·925	8·951	0·8633	15·22	137·74	62·96
3·82	118·385	30·044	3·940	9·032	0·8638	15·18	137·92	63·10
3·83	120·005	30·337	3·956	9·114	0·8644	15·14	138·10	63·24
3·84	121·645	30·632	3·971	9·196	0·8650	15·09	138·28	63·38
3·85	123·306	30·929	3·987	9·280	0·8655	15·05	138·46	63·52
3·86	124·986	31·229	4·002	9·363	0·8661	15·01	138·64	63·66
3·87	126·688	31·531	4·018	9·448	0·8667	14·98	138·82	63·79
3·88	128·411	31·836	4·033	9·533	0·8672	14·94	139·00	63·93
3·89	130·154	32·144	4·049	9·619	0·8678	14·90	139·17	64·07
3·90	131·919	32·454	4·065	9·706	0·8683	14·86	139·35	64·20
3·91	133·706	32·767	4·081	9·793	0·8689	14·82	139·52	64·34
3·92	135·514	33·082	4·096	9·881	0·8694	14·78	139·70	64·48
3·93	137·345	33·400	4·112	9·970	0·8700	14·74	139·87	64·61
3·94	139·198	33·720	4·128	10·060	0·8705	14·70	140·04	64·75
3·95	141·073	34·043	4·144	10·150	0·8710	14·66	140·21	64·88
3·96	142·971	34·369	4·160	10·241	0·8716	14·63	140·39	65·01
3·97	144·892	34·698	4·176	10·332	0·8721	14·59	140·56	65·15
3·98	146·836	35·029	4·192	10·425	0·8726	14·55	140·73	65·28
3·99	148·804	35·363	4·208	10·518	0·8731	14·51	140·90	65·41
4·00	150·796	35·700	4·224	10·612	0·8736	14·48	141·06	65·54

TABLE I B
ISENTROPIC FLOW OF DRY AIR WITH PRANDTL–MEYER EXPANSION ANGLES
(FOR EQUAL INCREMENTS OF SUPERSONIC FLOW DEFLEXION FROM 0 to 70°)

v	μ	θ	M	p_0/p	ρ_0/ρ	T_0/T	A/A^*	u/c
0.00	90.00	00.00	1.000	1.895	1.577	1.201	1.000	0.4095
0.20	76.75	13.45	1.027	1.957	1.614	1.213	1.001	0.4188
0.40	73.35	17.05	1.044	1.996	1.636	1.220	1.002	0.4243
0.60	70.99	19.61	1.058	2.029	1.656	1.225	1.003	0.4289
0.80	69.13	21.67	1.070	2.060	1.674	1.231	1.004	0.4330
1.00	67.56	23.44	1.082	2.090	1.691	1.236	1.005	0.4369
1.20	66.21	24.99	1.093	2.119	1.708	1.241	1.007	0.4404
1.40	65.00	26.40	1.103	2.146	1.724	1.245	1.008	0.4438
1.60	63.91	27.69	1.113	2.174	1.739	1.250	1.010	0.4471
1.80	62.91	28.89	1.123	2.200	1.754	1.254	1.012	0.4502
2.00	61.98	30.02	1.133	2.227	1.769	1.259	1.014	0.4532
2.20	61.13	31.07	1.142	2.253	1.784	1.263	1.016	0.4562
2.40	60.32	32.08	1.151	2.279	1.799	1.267	1.018	0.4590
2.60	59.56	33.04	1.160	2.305	1.813	1.271	1.020	0.4618
2.80	58.85	33.95	1.169	2.331	1.828	1.275	1.022	0.4645
3.00	58.17	34.83	1.177	2.356	1.842	1.279	1.024	0.4672
3.20	57.52	35.68	1.185	2.382	1.857	1.283	1.026	0.4698
3.40	56.90	36.50	1.194	2.408	1.871	1.287	1.029	0.4723
3.60	56.31	37.29	1.202	2.434	1.885	1.291	1.031	0.4748
3.80	55.74	38.06	1.210	2.460	1.899	1.295	1.033	0.4773
4.00	55.19	38.81	1.218	2.486	1.914	1.299	1.036	0.4797
4.20	54.66	39.54	1.226	2.512	1.928	1.303	1.038	0.4821
4.40	54.15	40.25	1.234	2.538	1.942	1.307	1.041	0.4845
4.60	53.66	40.94	1.241	2.564	1.956	1.311	1.044	0.4868
4.80	53.18	41.62	1.249	2.590	1.971	1.314	1.046	0.4891
5.00	52.72	42.28	1.257	2.617	1.985	1.318	1.049	0.4914
5.20	52.27	42.93	1.264	2.643	1.999	1.322	1.052	0.4936
5.40	51.84	43.56	1.272	2.670	2.014	1.326	1.055	0.4958
5.60	51.41	44.19	1.279	2.697	2.028	1.330	1.058	0.4980
5.80	51.00	44.80	1.287	2.725	2.043	1.334	1.061	0.5002
6.00	50.60	45.40	1.294	2.752	2.058	1.337	1.064	0.5023
6.20	50.21	45.99	1.301	2.779	2.072	1.341	1.067	0.5044
6.40	49.83	46.57	1.309	2.807	2.087	1.345	1.070	0.5065
6.60	49.45	47.15	1.316	2.835	2.102	1.349	1.073	0.5086
6.80	49.09	47.71	1.323	2.863	2.117	1.353	1.076	0.5107
7.00	48.73	48.27	1.330	2.892	2.132	1.357	1.080	0.5127
7.20	48.38	48.82	1.338	2.921	2.147	1.361	1.083	0.5148
7.40	48.04	49.36	1.345	2.950	2.162	1.364	1.086	0.5168
7.60	47.71	49.89	1.352	2.979	2.177	1.368	1.090	0.5188
7.80	47.38	50.42	1.359	3.008	2.192	1.372	1.093	0.5208
8.00	47.06	50.94	1.366	3.038	2.208	1.376	1.097	0.5227
8.20	46.75	51.45	1.373	3.068	2.223	1.380	1.100	0.5247
8.40	46.44	51.96	1.380	3.098	2.239	1.384	1.104	0.5266
8.60	46.13	52.47	1.387	3.129	2.255	1.388	1.108	0.5286
8.80	45.84	52.96	1.394	3.159	2.270	1.392	1.111	0.5305

TABLE I B .

ISENTROPIC FLOW OF DRY AIR WITH PRANDTL-MEYER EXPANSION ANGLES

(FOR EQUAL INCREMENTS OF SUPERSONIC FLOW DEFLEXION FROM 0 to 70°)

v	μ	θ	M	p_0/p	ρ_0/ρ	T_0/T	A/A^*	u/c
9.00	45.54	53.46	1.401	3.191	2.286	1.396	1.115	0.5324
9.20	45.26	53.94	1.408	3.222	2.302	1.399	1.119	0.5343
9.40	44.97	54.43	1.415	3.254	2.318	1.403	1.123	0.5361
9.60	44.70	54.90	1.422	3.286	2.335	1.407	1.127	0.5380
9.80	44.42	55.38	1.429	3.318	2.351	1.411	1.131	0.5398
10.00	44.15	55.85	1.436	3.351	2.368	1.415	1.135	0.5417
10.20	43.89	56.31	1.442	3.384	2.384	1.419	1.139	0.5435
10.40	43.63	56.77	1.449	3.417	2.401	1.423	1.143	0.5453
10.60	43.37	57.23	1.456	3.451	2.418	1.427	1.147	0.5471
10.80	43.12	57.68	1.463	3.485	2.435	1.431	1.152	0.5489
11.00	42.87	58.13	1.470	3.519	2.452	1.435	1.156	0.5507
11.20	42.62	58.58	1.477	3.554	2.469	1.439	1.160	0.5525
11.40	42.38	59.02	1.484	3.589	2.486	1.443	1.165	0.5543
11.60	42.14	59.46	1.490	3.624	2.504	1.448	1.169	0.5560
11.80	41.91	59.89	1.497	3.660	2.521	1.452	1.174	0.5578
12.00	41.67	60.33	1.504	3.697	2.539	1.456	1.178	0.5595
12.20	41.45	60.75	1.511	3.733	2.557	1.460	1.183	0.5613
12.40	41.22	61.18	1.518	3.770	2.575	1.464	1.188	0.5630
12.60	41.00	61.60	1.524	3.808	2.593	1.468	1.193	0.5647
12.80	40.78	62.02	1.531	3.846	2.612	1.472	1.197	0.5664
13.00	40.56	62.44	1.538	3.884	2.630	1.477	1.202	0.5681
13.20	40.34	62.86	1.545	3.923	2.649	1.481	1.207	0.5698
13.40	40.13	63.27	1.552	3.962	2.668	1.485	1.212	0.5715
13.60	39.92	63.68	1.558	4.001	2.687	1.489	1.217	0.5732
13.80	39.71	64.09	1.565	4.041	2.706	1.494	1.222	0.5749
14.00	39.51	64.49	1.572	4.082	2.725	1.498	1.228	0.5765
14.20	39.31	64.89	1.579	4.123	2.745	1.502	1.233	0.5782
14.40	39.11	65.29	1.585	4.164	2.764	1.506	1.238	0.5798
14.60	38.91	65.69	1.592	4.206	2.784	1.511	1.243	0.5815
14.80	38.71	66.09	1.599	4.249	2.804	1.515	1.249	0.5831
15.00	38.52	66.48	1.606	4.291	2.824	1.520	1.254	0.5847
15.20	38.33	66.87	1.613	4.335	2.845	1.524	1.260	0.5863
15.40	38.14	67.26	1.619	4.379	2.865	1.528	1.265	0.5880
15.60	37.95	67.65	1.626	4.423	2.886	1.533	1.271	0.5896
15.80	37.76	68.04	1.633	4.468	2.907	1.537	1.277	0.5912
16.00	37.58	68.42	1.640	4.514	2.928	1.542	1.283	0.5928
16.20	37.40	68.80	1.646	4.560	2.949	1.546	1.288	0.5944
16.40	37.22	69.18	1.653	4.606	2.970	1.551	1.294	0.5959
16.60	37.04	69.56	1.660	4.653	2.992	1.555	1.300	0.5975
16.80	36.87	69.93	1.667	4.701	3.014	1.560	1.306	0.5991
17.00	36.69	70.31	1.674	4.749	3.036	1.564	1.312	0.6007
17.20	36.52	70.68	1.680	4.798	3.058	1.569	1.319	0.6022
17.40	36.35	71.05	1.687	4.847	3.080	1.574	1.325	0.6038
17.60	36.18	71.42	1.694	4.897	3.103	1.578	1.331	0.6053
17.80	36.01	71.79	1.701	4.948	3.126	1.583	1.338	0.6069

11

TABLE I B

ISENTROPIC FLOW OF DRY AIR WITH PRANDTL-MEYER EXPANSION ANGLES

(FOR EQUAL INCREMENTS OF SUPERSONIC FLOW DEFLEXION FROM 0 to 70°)

ν	μ	θ	M	p_0/p	ρ_0/ρ	T_0/T	A/A^*	u/c
18.00	35.84	72.16	1.708	4.999	3.149	1.588	1.344	0.6084
18.20	35.68	72.52	1.715	5.051	3.172	1.592	1.351	0.6099
18.40	35.51	72.89	1.721	5.104	3.196	1.597	1.357	0.6115
18.60	35.35	73.25	1.728	5.157	3.219	1.602	1.364	0.6130
18.80	35.19	73.61	1.735	5.211	3.243	1.607	1.371	0.6145
19.00	35.03	73.97	1.742	5.265	3.267	1.611	1.377	0.6160
19.20	34.88	74.32	1.749	5.321	3.292	1.616	1.384	0.6175
19.40	34.72	74.68	1.756	5.377	3.316	1.621	1.391	0.6190
19.60	34.56	75.04	1.763	5.433	3.341	1.626	1.398	0.6205
19.80	34.41	75.39	1.770	5.490	3.366	1.631	1.405	0.6220
20.00	34.26	75.74	1.776	5.548	3.392	1.636	1.413	0.6235
20.20	34.11	76.09	1.783	5.607	3.417	1.641	1.420	0.6250
20.40	33.96	76.44	1.790	5.667	3.443	1.646	1.427	0.6264
20.60	33.81	76.79	1.797	5.727	3.469	1.651	1.435	0.6279
20.80	33.66	77.14	1.804	5.788	3.495	1.656	1.442	0.6294
21.00	33.51	77.49	1.811	5.850	3.522	1.661	1.450	0.6308
21.20	33.37	77.83	1.818	5.913	3.549	1.666	1.458	0.6323
21.40	33.23	78.17	1.825	5.976	3.576	1.671	1.465	0.6337
21.60	33.08	78.52	1.832	6.040	3.603	1.676	1.473	0.6352
21.80	32.94	78.86	1.839	6.105	3.631	1.681	1.481	0.6366
22.00	32.80	79.20	1.846	6.171	3.659	1.687	1.489	0.6381
22.20	32.66	79.54	1.853	6.238	3.687	1.692	1.497	0.6395
22.40	32.52	79.88	1.860	6.306	3.716	1.697	1.505	0.6409
22.60	32.38	80.22	1.867	6.374	3.744	1.702	1.514	0.6423
22.80	32.25	80.55	1.874	6.444	3.773	1.708	1.522	0.6438
23.00	32.11	80.89	1.881	6.514	3.803	1.713	1.531	0.6452
23.20	31.98	81.22	1.888	6.586	3.832	1.718	1.539	0.6466
23.40	31.84	81.56	1.895	6.658	3.862	1.724	1.548	0.6480
23.60	31.71	81.89	1.902	6.731	3.892	1.729	1.557	0.6494
23.80	31.58	82.22	1.910	6.805	3.923	1.735	1.565	0.6508
24.00	31.45	82.55	1.917	6.880	3.954	1.740	1.574	0.6522
24.20	31.32	82.88	1.924	6.957	3.985	1.746	1.583	0.6536
24.40	31.19	83.21	1.931	7.034	4.016	1.751	1.592	0.6550
24.60	31.06	83.54	1.938	7.112	4.048	1.757	1.602	0.6563
24.80	30.94	83.86	1.945	7.192	4.080	1.762	1.611	0.6577
25.00	30.81	84.19	1.952	7.272	4.113	1.768	1.620	0.6591
25.20	30.68	84.52	1.960	7.353	4.146	1.774	1.630	0.6605
25.40	30.56	84.84	1.967	7.436	4.179	1.779	1.640	0.6618
25.60	30.44	85.16	1.974	7.520	4.212	1.785	1.649	0.6632
25.80	30.31	85.49	1.981	7.605	4.246	1.791	1.659	0.6646
26.00	30.19	85.81	1.988	7.691	4.280	1.797	1.669	0.6659
26.20	30.07	86.13	1.996	7.778	4.315	1.803	1.679	0.6673
26.40	29.95	86.45	2.003	7.866	4.350	1.808	1.689	0.6686
26.60	29.83	86.77	2.010	7.956	4.385	1.814	1.700	0.6700
26.80	29.71	87.09	2.018	8.047	4.421	1.820	1.710	0.6713

TABLE I B

ISENTROPIC FLOW OF DRY AIR WITH PRANDTL–MEYER EXPANSION ANGLES

(FOR EQUAL INCREMENTS OF SUPERSONIC FLOW DEFLEXION FROM 0 to 70°)

ν	μ	θ	M	p_0/p	ρ_0/ρ	T_0/T	A/A^*	u/c
27.00	29.59	87.41	2.025	8.139	4.457	1.826	1.721	0.6726
27.20	29.48	87.72	2.032	8.233	4.493	1.832	1.731	0.6740
27.40	29.36	88.04	2.040	8.328	4.530	1.838	1.742	0.6753
27.60	29.24	88.36	2.047	8.424	4.567	1.844	1.753	0.6766
27.80	29.13	88.67	2.054	8.521	4.605	1.850	1.764	0.6779
28.00	29.01	88.99	2.062	8.620	4.643	1.857	1.775	0.6792
28.20	28.90	89.30	2.069	8.720	4.681	1.863	1.786	0.6806
28.40	28.79	89.61	2.077	8.822	4.720	1.869	1.798	0.6819
28.60	28.67	89.93	2.084	8.925	4.759	1.875	1.809	0.6832
28.80	28.56	90.24	2.092	9.029	4.799	1.882	1.821	0.6845
29.00	28.45	90.55	2.099	9.136	4.839	1.888	1.832	0.6858
29.20	28.34	90.86	2.107	9.243	4.880	1.894	1.844	0.6871
29.40	28.23	91.17	2.114	9.352	4.921	1.901	1.856	0.6884
29.60	28.12	91.48	2.122	9.463	4.962	1.907	1.868	0.6897
29.80	28.01	91.79	2.129	9.575	5.004	1.913	1.881	0.6909
30.00	27.90	92.10	2.137	9.689	5.046	1.920	1.893	0.6922
30.20	27.80	92.40	2.144	9.805	5.089	1.927	1.906	0.6935
30.40	27.69	92.71	2.152	9.922	5.133	1.933	1.918	0.6948
30.60	27.58	93.02	2.160	10.041	5.176	1.940	1.931	0.6960
30.80	27.48	93.32	2.167	10.162	5.221	1.946	1.944	0.6973
31.00	27.37	93.63	2.175	10.284	5.265	1.953	1.957	0.6986
31.20	27.27	93.93	2.183	10.408	5.311	1.960	1.971	0.6998
31.40	27.17	94.23	2.190	10.534	5.356	1.967	1.984	0.7011
31.60	27.06	94.54	2.198	10.662	5.403	1.974	1.998	0.7023
31.80	26.96	94.84	2.206	10.792	5.449	1.980	2.011	0.7036
32.00	26.86	95.14	2.214	10.924	5.497	1.987	2.025	0.7048
32.20	26.76	95.44	2.221	11.058	5.545	1.994	2.039	0.7061
32.40	26.65	95.75	2.229	11.193	5.593	2.001	2.053	0.7073
32.60	26.55	96.05	2.237	11.331	5.642	2.008	2.068	0.7086
32.80	26.45	96.35	2.245	11.471	5.692	2.015	2.082	0.7098
33.00	26.35	96.65	2.253	11.613	5.742	2.023	2.097	0.7110
33.20	26.25	96.95	2.261	11.757	5.792	2.030	2.112	0.7123
33.40	26.16	97.24	2.268	11.903	5.844	2.037	2.127	0.7135
33.60	26.06	97.54	2.276	12.052	5.896	2.044	2.142	0.7147
33.80	25.96	97.84	2.284	12.202	5.948	2.051	2.157	0.7159
34.00	25.86	98.14	2.292	12.355	6.001	2.059	2.173	0.7171
34.20	25.77	98.43	2.300	12.511	6.055	2.066	2.189	0.7184
34.40	25.67	98.73	2.308	12.668	6.109	2.074	2.205	0.7196
34.60	25.58	99.02	2.316	12.828	6.164	2.081	2.221	0.7208
34.80	25.48	99.32	2.324	12.991	6.220	2.089	2.237	0.7220
35.00	25.39	99.61	2.333	13.156	6.276	2.096	2.254	0.7232
35.20	25.29	99.91	2.341	13.324	6.333	2.104	2.270	0.7244
35.40	25.20	100.20	2.349	13.494	6.390	2.112	2.287	0.7256
35.60	25.10	100.50	2.357	13.667	6.448	2.119	2.304	0.7267
35.80	25.01	100.79	2.365	13.842	6.507	2.127	2.321	0.7279

TABLE I B

ISENTROPIC FLOW OF DRY AIR WITH PRANDTL-MEYER EXPANSION ANGLES

(FOR EQUAL INCREMENTS OF SUPERSONIC FLOW DEFLEXION FROM 0 to 70°)

v	μ	θ	M	p_0/p	ρ_0/ρ	T_0/T	A/A^*	u/c
36.00	24.92	101.08	2.373	14.021	6.567	2.135	2.339	0.7291
36.20	24.83	101.37	2.382	14.202	6.627	2.143	2.357	0.7303
36.40	24.74	101.66	2.390	14.386	6.688	2.151	2.374	0.7315
36.60	24.65	101.96	2.398	14.572	6.750	2.159	2.393	0.7327
36.80	24.55	102.25	2.406	14.762	6.813	2.167	2.411	0.7338
37.00	24.46	102.54	2.415	14.955	6.876	2.175	2.429	0.7350
37.20	24.37	102.83	2.423	15.151	6.940	2.183	2.448	0.7362
37.40	24.28	103.12	2.431	15.350	7.005	2.191	2.467	0.7373
37.60	24.20	103.40	2.440	15.552	7.071	2.200	2.486	0.7385
37.80	24.11	103.69	2.448	15.757	7.137	2.208	2.506	0.7396
38.00	24.02	103.98	2.457	15.966	7.204	2.216	2.525	0.7408
38.20	23.93	104.27	2.465	16.178	7.272	2.225	2.545	0.7419
38.40	23.84	104.56	2.474	16.393	7.341	2.233	2.565	0.7431
38.60	23.76	104.84	2.482	16.612	7.411	2.242	2.586	0.7442
38.80	23.67	105.13	2.491	16.835	7.481	2.250	2.607	0.7454
39.00	23.58	105.42	2.499	17.061	7.553	2.259	2.627	0.7465
39.20	23.50	105.70	2.508	17.290	7.625	2.268	2.649	0.7477
39.40	23.41	105.99	2.517	17.524	7.699	2.276	2.670	0.7488
39.60	23.33	106.27	2.525	17.761	7.773	2.285	2.692	0.7499
39.80	23.24	106.56	2.534	18.002	7.848	2.294	2.714	0.7510
40.00	23.16	106.84	2.543	18.248	7.924	2.303	2.736	0.7522
40.20	23.07	107.13	2.552	18.497	8.001	2.312	2.758	0.7533
40.40	22.99	107.41	2.560	18.750	8.079	2.321	2.781	0.7544
40.60	22.91	107.69	2.569	19.008	8.158	2.330	2.804	0.7555
40.80	22.82	107.98	2.578	19.269	8.238	2.339	2.827	0.7566
41.00	22.74	108.26	2.587	19.535	8.319	2.348	2.851	0.7577
41.20	22.66	108.54	2.596	19.806	8.400	2.358	2.875	0.7589
41.40	22.58	108.82	2.605	20.081	8.483	2.367	2.899	0.7600
41.60	22.49	109.11	2.614	20.361	8.568	2.377	2.923	0.7611
41.80	22.41	109.39	2.623	20.645	8.653	2.386	2.948	0.7622
42.00	22.33	109.67	2.632	20.935	8.739	2.396	2.973	0.7633
42.20	22.25	109.95	2.641	21.229	8.826	2.405	2.999	0.7644
42.40	22.17	110.23	2.650	21.528	8.915	2.415	3.024	0.7654
42.60	22.09	110.51	2.659	21.832	9.004	2.425	3.051	0.7665
42.80	22.01	110.79	2.668	22.142	9.095	2.434	3.077	0.7676
43.00	21.93	111.07	2.677	22.456	9.187	2.444	3.104	0.7687
43.20	21.85	111.35	2.687	22.777	9.280	2.454	3.131	0.7698
43.40	21.77	111.63	2.696	23.102	9.375	2.464	3.158	0.7708
43.60	21.70	111.90	2.705	23.433	9.470	2.474	3.186	0.7719
43.80	21.62	112.18	2.714	23.770	9.567	2.485	3.214	0.7730
44.00	21.54	112.46	2.724	24.113	9.665	2.495	3.243	0.7741
44.20	21.46	112.74	2.733	24.462	9.765	2.505	3.271	0.7751
44.40	21.38	113.02	2.742	24.817	9.865	2.516	3.301	0.7762
44.60	21.31	113.29	2.752	25.178	9.968	2.526	3.330	0.7772
44.80	21.23	113.57	2.761	25.545	10.071	2.537	3.360	0.7783

TABLE I B
ISENTROPIC FLOW OF DRY AIR WITH PRANDTL-MEYER EXPANSION ANGLES

(FOR EQUAL INCREMENTS OF SUPERSONIC FLOW DEFLEXION FROM 0 to 70°)

ν	μ	θ	M	p_0/p	ρ_0/ρ	T_0/T	A/A^*	u/c
45.00	21.15	113.85	2.771	25.919	10.176	2.547	3.391	0.7794
45.20	21.08	114.12	2.780	26.300	10.282	2.558	3.421	0.7804
45.40	21.00	114.40	2.790	26.687	10.390	2.569	3.453	0.7815
45.60	20.93	114.67	2.800	27.081	10.499	2.579	3.484	0.7825
45.80	20.85	114.95	2.809	27.482	10.609	2.590	3.516	0.7835
46.00	20.78	115.22	2.819	27.890	10.722	2.601	3.549	0.7846
46.20	20.70	115.50	2.829	28.306	10.835	2.612	3.582	0.7856
46.40	20.63	115.77	2.839	28.729	10.950	2.624	3.615	0.7867
46.60	20.55	116.05	2.848	29.159	11.067	2.635	3.649	0.7877
46.80	20.48	116.32	2.858	29.598	11.185	2.646	3.683	0.7887
47.00	20.41	116.59	2.868	30.044	11.305	2.657	3.717	0.7898
47.20	20.33	116.87	2.878	30.498	11.427	2.669	3.753	0.7908
47.40	20.26	117.14	2.888	30.961	11.550	2.681	3.788	0.7918
47.60	20.19	117.41	2.898	31.432	11.675	2.692	3.824	0.7928
47.80	20.11	117.69	2.908	31.911	11.802	2.704	3.861	0.7938
48.00	20.04	117.96	2.918	32.400	11.930	2.716	3.898	0.7948
48.20	19.97	118.23	2.928	32.897	12.061	2.728	3.935	0.7959
48.40	19.90	118.50	2.938	33.404	12.193	2.740	3.973	0.7969
48.60	19.83	118.77	2.948	33.920	12.327	2.752	4.012	0.7979
48.80	19.75	119.05	2.959	34.445	12.462	2.764	4.051	0.7989
49.00	19.68	119.32	2.969	34.980	12.600	2.776	4.091	0.7999
49.20	19.61	119.59	2.979	35.526	12.740	2.789	4.131	0.8009
49.40	19.54	119.86	2.990	36.081	12.881	2.801	4.172	0.8019
49.60	19.47	120.13	3.000	36.647	13.025	2.814	4.213	0.8029
49.80	19.40	120.40	3.010	37.223	13.171	2.826	4.255	0.8038
50.00	19.33	120.67	3.021	37.811	13.319	2.839	4.297	0.8048
50.20	19.26	120.94	3.031	38.409	13.468	2.852	4.340	0.8058
50.40	19.19	121.21	3.042	39.019	13.621	2.865	4.384	0.8068
50.60	19.12	121.48	3.053	39.640	13.775	2.878	4.428	0.8078
50.80	19.05	121.75	3.063	40.273	13.931	2.891	4.473	0.8088
51.00	18.98	122.02	3.074	40.919	14.090	2.904	4.519	0.8097
51.20	18.92	122.28	3.085	41.576	14.251	2.917	4.565	0.8107
51.40	18.85	122.55	3.096	42.247	14.414	2.931	4.612	0.8117
51.60	18.78	122.82	3.106	42.930	14.580	2.944	4.659	0.8126
51.80	18.71	123.09	3.117	43.626	14.748	2.958	4.707	0.8136
52.00	18.64	123.36	3.128	44.336	14.919	2.972	4.756	0.8146
52.20	18.58	123.62	3.139	45.060	15.092	2.986	4.806	0.8155
52.40	18.51	123.89	3.150	45.798	15.268	3.000	4.856	0.8165
52.60	18.44	124.16	3.161	46.550	15.446	3.014	4.907	0.8174
52.80	18.37	124.43	3.172	47.318	15.627	3.028	4.959	0.8184
53.00	18.31	124.69	3.184	48.100	15.811	3.042	5.011	0.8193
53.20	18.24	124.96	3.195	48.897	15.998	3.057	5.065	0.8203
53.40	18.17	125.23	3.206	49.711	16.187	3.071	5.119	0.8212
53.60	18.11	125.49	3.217	50.541	16.379	3.086	5.173	0.8221
53.80	18.04	125.76	3.229	51.387	16.574	3.100	5.229	0.8231

TABLE I B

ISENTROPIC FLOW OF DRY AIR WITH PRANDTL-MEYER EXPANSION ANGLES
(FOR EQUAL INCREMENTS OF SUPERSONIC FLOW DEFLEXION FROM 0 to 70°)

v	μ	θ	M	p_0/p	ρ_0/ρ	T_0/T	A/A^*	u/c
54.00	17.98	126.02	3.240	52.250	16.772	3.115	5.286	0.8240
54.20	17.91	126.29	3.252	53.130	16.973	3.130	5.343	0.8250
54.40	17.85	126.55	3.263	54.028	17.177	3.145	5.401	0.8259
54.60	17.78	126.82	3.275	54.944	17.384	3.161	5.460	0.8268
54.80	17.72	127.08	3.286	55.879	17.594	3.176	5.520	0.8277
55.00	17.65	127.35	3.298	56.833	17.808	3.191	5.581	0.8287
55.20	17.59	127.61	3.310	57.806	18.024	3.207	5.642	0.8296
55.40	17.52	127.88	3.321	58.799	18.244	3.223	5.705	0.8305
55.60	17.46	128.14	3.333	59.812	18.468	3.239	5.768	0.8314
55.80	17.39	128.41	3.345	60.846	18.695	3.255	5.833	0.8323
56.00	17.33	128.67	3.357	61.901	18.926	3.271	5.898	0.8332
56.20	17.27	128.93	3.369	62.978	19.160	3.287	5.965	0.8341
56.40	17.20	129.20	3.381	64.077	19.397	3.303	6.032	0.8350
56.60	17.14	129.46	3.393	65.200	19.639	3.320	6.101	0.8359
56.80	17.08	129.72	3.405	66.345	19.884	3.337	6.171	0.8368
57.00	17.01	129.99	3.417	67.515	20.133	3.353	6.241	0.8377
57.20	16.95	130.25	3.430	68.709	20.387	3.370	6.313	0.8386
57.40	16.89	130.51	3.442	69.928	20.644	3.387	6.386	0.8395
57.60	16.83	130.77	3.454	71.173	20.905	3.405	6.460	0.8404
57.80	16.76	131.04	3.467	72.444	21.171	3.422	6.535	0.8413
58.00	16.70	131.30	3.479	73.742	21.440	3.439	6.611	0.8422
58.20	16.64	131.56	3.492	75.068	21.714	3.457	6.689	0.8431
58.40	16.58	131.82	3.505	76.422	21.993	3.475	6.767	0.8439
58.60	16.52	132.08	3.517	77.806	22.276	3.493	6.847	0.8448
58.80	16.46	132.34	3.530	79.218	22.563	3.511	6.929	0.8457
59.00	16.40	132.60	3.543	80.662	22.856	3.529	7.011	0.8466
59.20	16.33	132.87	3.556	82.136	23.153	3.548	7.095	0.8474
59.40	16.27	133.13	3.569	83.643	23.455	3.566	7.180	0.8483
59.60	16.21	133.39	3.582	85.182	23.761	3.585	7.267	0.8491
59.80	16.15	133.65	3.595	86.755	24.073	3.604	7.355	0.8500
60.00	16.09	133.91	3.608	88.363	24.390	3.623	7.444	0.8509
60.20	16.03	134.17	3.621	90.006	24.713	3.642	7.535	0.8517
60.40	15.97	134.43	3.634	91.685	25.041	3.661	7.627	0.8526
60.60	15.91	134.69	3.648	93.401	25.374	3.681	7.721	0.8534
60.80	15.85	134.95	3.661	95.155	25.712	3.701	7.816	0.8543
61.00	15.79	135.21	3.674	96.948	26.057	3.721	7.913	0.8551
61.20	15.73	135.47	3.688	98.781	26.407	3.741	8.012	0.8560
61.40	15.67	135.73	3.702	100.656	26.763	3.761	8.112	0.8568
61.60	15.61	135.99	3.715	102.572	27.125	3.781	8.213	0.8576
61.80	15.56	136.24	3.729	104.532	27.494	3.802	8.317	0.8585
62.00	15.50	136.50	3.743	106.536	27.869	3.823	8.422	0.8593
62.20	15.44	136.76	3.757	108.585	28.250	3.844	8.529	0.8601
62.40	15.38	137.02	3.771	110.681	28.637	3.865	8.638	0.8610
62.60	15.32	137.28	3.785	112.825	29.031	3.886	8.748	0.8618
62.80	15.26	137.54	3.799	115.018	29.433	3.908	8.861	0.8626

TABLE I B

ISENTROPIC FLOW OF DRY AIR WITH PRANDTL-MEYER EXPANSION ANGLES
(FOR EQUAL INCREMENTS OF SUPERSONIC FLOW DEFLEXION FROM 0 to 70°)

ν	μ	θ	M	p_0/p	ρ_0/ρ	T_0/T	A/A^*	u/c
63.00	15.20	137.80	3.813	117.262	29.841	3.930	8.975	0.8634
63.20	15.15	138.05	3.827	119.557	30.256	3.952	9.091	0.8643
63.40	15.09	138.31	3.842	121.905	30.678	3.974	9.209	0.8651
63.60	15.03	138.57	3.856	124.308	31.108	3.996	9.330	0.8659
63.80	14.97	138.83	3.870	126.767	31.545	4.019	9.452	0.8667
64.00	14.92	139.08	3.885	129.283	31.990	4.041	9.576	0.8675
64.20	14.86	139.34	3.900	131.859	32.443	4.064	9.703	0.8683
64.40	14.80	139.60	3.914	134.495	32.904	4.087	9.832	0.8691
64.60	14.74	139.86	3.929	137.193	33.373	4.111	9.963	0.8699
64.80	14.69	140.11	3.944	139.955	33.851	4.134	10.096	0.8707
65.00	14.63	140.37	3.959	142.783	34.337	4.158	10.232	0.8715
65.20	14.57	140.63	3.974	145.678	34.832	4.182	10.370	0.8723
65.40	14.52	140.88	3.989	148.642	35.336	4.207	10.510	0.8731
65.60	14.46	141.14	4.004	151.678	35.848	4.231	10.653	0.8739
65.80	14.41	141.39	4.020	154.787	36.371	4.256	10.798	0.8747
66.00	14.35	141.65	4.035	157.971	36.902	4.281	10.947	0.8754
66.20	14.29	141.91	4.051	161.232	37.444	4.306	11.097	0.8762
66.40	14.24	142.16	4.066	164.572	37.995	4.331	11.251	0.8770
66.60	14.18	142.42	4.082	167.995	38.557	4.357	11.407	0.8778
66.80	14.13	142.67	4.097	171.501	39.128	4.383	11.566	0.8785
67.00	14.07	142.93	4.113	175.093	39.711	4.409	11.728	0.8793
67.20	14.02	143.18	4.129	178.774	40.304	4.436	11.892	0.8801
67.40	13.96	143.44	4.145	182.546	40.908	4.462	12.060	0.8809
67.60	13.90	143.70	4.161	186.412	41.524	4.489	12.231	0.8816
67.80	13.85	143.95	4.177	190.374	42.151	4.516	12.405	0.8824
68.00	13.79	144.21	4.194	194.436	42.790	4.544	12.583	0.8831
68.20	13.74	144.46	4.210	198.599	43.441	4.572	12.763	0.8839
68.40	13.69	144.71	4.227	202.868	44.105	4.600	12.947	0.8846
68.60	13.63	144.97	4.243	207.244	44.781	4.628	13.134	0.8854
68.80	13.58	145.22	4.260	211.732	45.470	4.657	13.325	0.8861
69.00	13.52	145.48	4.277	216.334	46.172	4.685	13.520	0.8869
69.20	13.47	145.73	4.294	221.053	46.888	4.715	13.718	0.8876
69.40	13.41	145.99	4.310	225.894	47.617	4.744	13.919	0.8884
69.60	13.36	146.24	4.328	230.859	48.361	4.774	14.125	0.8891
69.80	13.31	146.49	4.345	235.953	49.119	4.804	14.335	0.8898
70.00	13.25	146.75	4.362	241.179	49.892	4.834	14.548	0.8906

TABLE II

FRICTIONLESS FLOW OF DRY AIR IN A CONSTANT AREA DUCT WITH HEAT TRANSFE

M	p/p^*	T/T^*	T_0/T_0^* $\left(1-\dfrac{q}{c_p T_0^*}\right)$	u/u^* (ρ^*/ρ)	p_0/p_0^*	$\Delta S/c_v$
0·00	2·4030	0·00000	0·00000	0·00000	1·2682	– inf.
0·01	2·4027	0·00058	0·00048	0·00024	1·2682	−10·816
0·02	2·4017	0·00231	0·00192	0·00096	1·2679	−8·872
0·03	2·4000	0·00518	0·00432	0·00216	1·2674	−7·736
0·04	2·3976	0·00920	0·00766	0·00384	1·2668	−6·931
0·05	2·3946	0·01434	0·01194	0·00599	1·2660	−6·308
0·06	2·3909	0·02058	0·01714	0·00861	1·2651	−5·800
0·07	2·3866	0·02791	0·02325	0·01169	1·2639	−5·371
0·08	2·3816	0·03630	0·03025	0·01524	1·2626	−5·002
0·09	2·3760	0·04573	0·03812	0·01925	1·2611	−4·677
0·10	2·3698	0·05616	0·04683	0·02370	1·2595	−4·388
0·11	2·3629	0·06756	0·05636	0·02859	1·2577	−4·127
0·12	2·3554	0·07989	0·06669	0·03392	1·2557	−3·891
0·13	2·3473	0·09312	0·07777	0·03967	1·2536	−3·674
0·14	2·3387	0·10720	0·08958	0·04584	1·2514	−3·475
0·15	2·3295	0·12209	0·10208	0·05241	1·2490	−3·291
0·16	2·3197	0·13775	0·11524	0·05938	1·2464	−3·120
0·17	2·3094	0·15413	0·12903	0·06674	1·2437	−2·961
0·18	2·2985	0·17117	0·14340	0·07447	1·2409	−2·812
0·19	2·2872	0·18884	0·15832	0·08257	1·2380	−2·672
0·20	2·2753	0·20708	0·17374	0·09101	1·2349	−2·541
0·21	2·2630	0·22584	0·18964	0·09980	1·2317	−2·417
0·22	2·2502	0·24507	0·20596	0·10891	1·2284	−2·300
0·23	2·2370	0·26471	0·22267	0·11834	1·2250	−2·189
0·24	2·2233	0·28473	0·23973	0·12806	1·2215	−2·084
0·25	2·2093	0·30506	0·25709	0·13808	1·2179	−1·985
0·26	2·1948	0·32565	0·27473	0·14837	1·2143	−1·891
0·27	2·1800	0·34646	0·29259	0·15892	1·2105	−1·801
0·28	2·1649	0·36744	0·31065	0·16973	1·2067	−1·716
0·29	2·1494	0·38853	0·32885	0·18076	1·2027	−1·635
0·30	2·1336	0·40970	0·34717	0·19202	1·1988	−1·557
0·31	2·1175	0·43089	0·36557	0·20349	1·1947	−1·483
0·32	2·1011	0·45207	0·38402	0·21516	1·1906	−1·413
0·33	2·0845	0·47319	0·40248	0·22700	1·1865	−1·346
0·34	2·0677	0·49421	0·42091	0·23902	1·1823	−1·282
0·35	2·0506	0·51509	0·43929	0·25120	1·1781	−1·220
0·36	2·0333	0·53580	0·45759	0·26351	1·1739	−1·161
0·37	2·0158	0·55630	0·47577	0·27597	1·1696	−1·105
0·38	1·9982	0·57655	0·49382	0·28854	1·1653	−1·052
0·39	1·9804	0·59653	0·51170	0·30122	1·1610	−1·000
0·40	1·9625	0·61620	0·52940	0·31399	1·1567	−0·951
0·41	1·9444	0·63555	0·54688	0·32686	1·1524	−0·903
0·42	1·9263	0·65453	0·56413	0·33979	1·1481	−0·858
0·43	1·9080	0·67314	0·58113	0·35279	1·1438	−0·815
0·44	1·8897	0·69135	0·59785	0·36585	1·1395	−0·774

TABLE II

FRICTIONLESS FLOW OF DRY AIR IN A CONSTANT AREA DUCT WITH HEAT TRANSFER

M	p/p^*	T/T^*	T_0/T_0^* $\left(1-\dfrac{q}{C_p T_0^*}\right)$	u/u^* (ρ^*/ρ)	p_0/p_0^*	$\Delta S/c_V$
0·45	1·8713	0·70914	0·61429	0·37895	1·1352	−0·7348
0·46	1·8529	0·72649	0·63043	0·39208	1·1309	−0·6969
0·47	1·8345	0·74338	0·64625	0·40523	1·1267	−0·6606
0·48	1·8160	0·75981	0·66174	0·41840	1·1225	−0·6258
0·49	1·7975	0·77576	0·67690	0·43158	1·1183	−0·5295
0·50	1·7790	0·79122	0·69170	0·44475	1·1141	−0·5607
0·51	1·7605	0·80618	0·70615	0·45792	1·1100	−0·5302
0·52	1·7421	0·82064	0·72023	0·47106	1·1060	−0·5010
0·53	1·7237	0·83458	0·73393	0·48418	1·1019	−0·4731
0·54	1·7053	0·84801	0·74727	0·49727	1·0980	−0·4464
0·55	1·6870	0·86092	0·76022	0·51032	1·0940	−0·4209
0·56	1·6688	0·87331	0·77278	0·52333	1·0902	−0·3964
0·57	1·6506	0·88518	0·78496	0·53628	1·0864	−0·3731
0·58	1·6325	0·89653	0·79676	0·54918	1·0826	−0·3507
0·59	1·6145	0·90736	0·80816	0·56201	1·0789	−0·3294
0·60	1·5966	0·91768	0·81918	0·57477	1·0753	−0·3091
0·61	1·5788	0·92748	0·82982	0·58747	1·0717	−0·2897
0·62	1·5611	0·93678	0·84007	0·60008	1·0683	−0·2711
0·63	1·5435	0·94557	0·84993	0·61262	1·0648	−0·2535
0·64	1·5260	0·95387	0·85942	0·62506	1·0615	−0·2366
0·65	1·5087	0·96168	0·86854	0·63742	1·0582	−0·2206
0·66	1·4915	0·96900	0·87728	0·64969	1·0551	−0·2053
0·67	1·4744	0·97585	0·88566	0·66186	1·0519	−0·1908
0·68	1·4575	0·98224	0·89368	0·67393	1·0489	−0·1770
0·69	1·4407	0·98817	0·90134	0·68591	1·0460	−0·1638
0·70	1·4240	0·99365	0·90866	0·69777	1·0431	−0·1514
0·71	1·4075	0·99869	0·91563	0·70953	1·0403	−0·1396
0·72	1·3912	1·00330	0·92226	0·72119	1·0377	−0·1285
0·73	1·3750	1·00749	0·92857	0·73273	1·0350	−0·1178
0·74	1·3589	1·01127	0·93455	0·74416	1·0325	−0·1079
0·75	1·3431	1·01465	0·94021	0·75548	1·0301	−0·0985
0·76	1·3274	1·01765	0·94556	0·76668	1·0278	−0·0896
0·77	1·3118	1·02027	0·95061	0·77776	1·0255	−0·0812
0·78	1·2964	1·02252	0·95537	0·78873	1·0234	−0·0734
0·79	1·2812	1·02441	0·95983	0·79959	1·0213	−0·0660
0·80	1·2661	1·02596	0·96402	0·81032	1·0193	−0·0591
0·81	1·2512	1·02718	0·96793	0·82093	1·0175	−0·0527
0·82	1·2365	1·02807	0·97158	0·83143	1·0157	−0·0467
0·83	1·2220	1·02864	0·97497	0·84180	1·0140	−0·0412
0·84	1·2076	1·02891	0·97811	0·85206	1·0124	−0·0360
0·85	1·1933	1·02889	0·98101	0·86219	1·0109	−0·0313
0·86	1·1793	1·02859	0·98367	0·87221	1·0095	−0·0269
0·87	1·1654	1·02801	0·98610	0·88210	1·0082	−0·0229
0·88	1·1517	1·02717	0·98831	0·89188	1·0070	−0·0193
0·89	1·1382	1·02608	0·99030	0·90153	1·0059	−0·0160

TABLE II

FRICTIONLESS FLOW OF DRY AIR IN A CONSTANT AREA DUCT WITH HEAT TRANSFE

M	p/p^*	T/T^*	T_0/T_0^* $\left(1-\dfrac{q}{C_p T_0^*}\right)$	u/u^* (ρ^*/ρ)	p_0/p_0^*	$\Delta S/c_v$
0.90	1·1248	1·02474	0·99209	0·91107	1·0049	-0·0131
0.91	1·1116	1·02317	0·99368	0·92048	1·0039	-0·0105
0.92	1·0985	1·02138	0·99507	0·92978	1·0031	-0·0082
0.93	1·0856	1·01937	0·99627	0·93896	1·0024	-0·0062
0.94	1·0729	1·01715	0·99730	0·94803	1·0017	-0·0045
0.95	1·0604	1·01474	0·99815	0·95698	1·0012	-0·0031
0.96	1·0480	1·01214	0·99883	0·96581	1·0008	-0·0020
0.97	1·0357	1·00936	0·99935	0·97453	1·0004	-0·0011
0.98	1·0237	1·00640	0·99972	0·98313	1·0002	-0·0005
0.99	1·0118	1·00328	0·99993	0·99162	1·0000	-0·0001
1·00	1·0000	1·0000	1·00000	1·0000	1·0000	0·000
1·01	0·9884	0·9966	0·99993	1·0083	1·0000	-0·0001
1·02	0·9770	0·9930	0·99973	1·0164	1·0002	-0·0005
1·03	0·9657	0·9893	0·99940	1·0245	1·0004	-0·0010
1·04	0·9545	0·9855	0·99895	1·0324	1·0008	-0·0018
1·05	0·9435	0·9815	0·99838	1·0402	1·0012	-0·0027
1·06	0·9327	0·9774	0·99770	1·0480	1·0017	-0·0040
1·07	0·9220	0·9733	0·99691	1·0556	1·0024	-0·0053
1·08	0·9114	0·9690	0·99602	1·0631	1·0031	-0·0068
1·09	0·9010	0·9646	0·99503	1·0705	1·0039	-0·0086
1·10	0·8908	0·9601	0·99394	1·0778	1·0049	-0·0105
1·11	0·8807	0·9556	0·99276	1·0851	1·0059	-0·0125
1·12	0·8707	0·9509	0·99150	1·0922	1·0070	-0·0148
1·13	0·8608	0·9462	0·99016	1·0992	1·0082	-0·0171
1·14	0·8511	0·9414	0·98874	1·1061	1·0095	-0·0197
1·15	0·8415	0·9366	0·98724	1·1129	1·0109	-0·0224
1·16	0·8321	0·9317	0·98568	1·1197	1·0124	-0·0252
1·17	0·8228	0·9267	0·98405	1·1263	1·0140	-0·0282
1·18	0·8136	0·9217	0·98235	1·1329	1·0157	-0·0313
1·19	0·8045	0·9166	0·98059	1·1393	1·0175	-0·0345
1·20	0·7956	0·9115	0·97878	1·1457	1·0194	-0·0378
1·21	0·7868	0·9064	0·97691	1·1520	1·0214	-0·0413
1·22	0·7781	0·9012	0·97499	1·1581	1·0235	-0·0449
1·23	0·7696	0·8960	0·97302	1·1643	1·0256	-0·0486
1·24	0·7611	0·8907	0·97101	1·1703	1·0279	-0·0524
1·25	0·7528	0·8854	0·96895	1·1762	1·0303	-0·0563
1·26	0·7446	0·8801	0·96685	1·1821	1·0328	-0·0603
1·27	0·7365	0·8748	0·96472	1·1878	1·0353	-0·0644
1·28	0·7285	0·8695	0·96255	1·1935	1·0380	-0·0686
1·29	0·7206	0·8641	0·96034	1·1991	1·0408	-0·0729
1·30	0·7128	0·8587	0·95810	1·2047	1·0436	-0·0773

TABLE II

FRICTIONLESS FLOW OF DRY AIR IN A CONSTANT AREA DUCT WITH HEAT TRANSFER

M	p/p^*	T/T^*	T_0/T_0^* $\left(1-\dfrac{q}{C_p T_0^*}\right)$	u/u^* (ρ^*/ρ)	p_0/p_0^*	$\Delta S/c_v$
1·31	0·7052	0·8534	0·95584	1·2101	1·0466	-0·0817
1·32	0·6976	0·8480	0·95355	1·2155	1·0496	-0·0862
1·33	0·6902	0·8426	0·95123	1·2208	1·0528	-0·0908
1·34	0·6828	0·8372	0·94889	1·2261	1·0560	-0·0956
1·35	0·6756	0·8318	0·94653	1·2312	1·0594	-0·1003
1·36	0·6684	0·8264	0·94415	1·2363	1·0628	-0·1052
1·37	0·6614	0·8210	0·94175	1·2414	1·0663	-0·1101
1·38	0·6544	0·8156	0·93933	1·2463	1·0700	-0·1150
1·39	0·6476	0·8102	0·93690	1·2512	1·0737	-0·1201
1·40	0·6408	0·8049	0·93446	1·2560	1·0776	-0·1252
1·41	0·6342	0·7995	0·93200	1·2608	1·0815	-0·1304
1·42	0·6276	0·7942	0·92954	1·2654	1·0855	-0·1356
1·43	0·6211	0·7888	0·92706	1·2701	1·0897	-0·1409
1·44	0·6147	0·7835	0·92458	1·2746	1·0939	-0·1462
1·45	0·6084	0·7782	0·92209	1·2791	1·0982	-0·1515
1·46	0·6022	0·7729	0·91959	1·2836	1·1026	-0·1570
1·47	0·5960	0·7676	0·91709	1·2879	1·1072	-0·1624
1·48	0·5900	0·7624	0·91459	1·2923	1·1118	-0·1680
1·49	0·5840	0·7572	0·91208	1·2965	1·1165	-0·1735
1·50	0·5781	0·7519	0·90957	1·3007	1·1214	-0·1791
1·51	0·5723	0·7467	0·90706	1·3049	1·1263	-0·1848
1·52	0·5665	0·7416	0·90455	1·3089	1·1313	-0·1905
1·53	0·5609	0·7364	0·90204	1·3130	1·1365	-0·1962
1·54	0·5553	0·7313	0·89954	1·3170	1·1417	-0·2020
1·55	0·5498	0·7262	0·89703	1·3209	1·1471	-0·2078
1·56	0·5444	0·7211	0·89453	1·3248	1·1525	-0·2136
1·57	0·5390	0·7161	0·89203	1·3286	1·1581	-0·2194
1·58	0·5337	0·7111	0·88954	1·3324	1·1637	-0·2253
1·59	0·5285	0·7061	0·88705	1·3361	1·1695	-0·2313
1·60	0·5233	0·7011	0·88457	1·3397	1·1753	-0·2372
1·61	0·5183	0·6962	0·88210	1·3434	1·1813	-0·2431
1·62	0·5132	0·6913	0·87963	1·3469	1·1874	-0·2492
1·63	0·5083	0·6864	0·87717	1·3505	1·1935	-0·2552
1·64	0·5034	0·6816	0·87472	1·3540	1·1998	-0·2612
1·65	0·4986	0·6768	0·87227	1·3574	1·2062	-0·2673
1·66	0·4938	0·6720	0·86984	1·3608	1·2127	-0·2734
1·67	0·4891	0·6672	0·86741	1·3641	1·2193	-0·2795
1·68	0·4845	0·6625	0·86499	1·3674	1·2260	-0·2856
1·69	0·4799	0·6578	0·86259	1·3707	1·2329	-0·2918
1·70	0·4754	0·6532	0·86019	1·3739	1·2398	-0·2979
1·71	0·4709	0·6485	0·85780	1·3771	1·2468	-0·3041
1·72	0·4665	0·6439	0·85543	1·3802	1·2540	-0·3103
1·73	0·4622	0·6394	0·85307	1·3833	1·2612	-0·3165
1·74	0·4579	0·6348	0·85071	1·3864	1·2686	-0·3227
1·75	0·4537	0·6303	0·84837	1·3894	1·2761	-0·3290

TABLE II

FRICTIONLESS FLOW OF DRY AIR IN A CONSTANT AREA DUCT WITH HEAT TRANSFE

M	p/p^*	T/T^*	T_0/T_0^* $\left(1-\dfrac{q}{C_p T_0^*}\right)$	u/u^* (ρ^*/ρ)	p_0/p_0^*	$\Delta S/c_v$
1·76	0·4495	0·6259	0·84604	1·3924	1·2837	−0·3352
1·77	0·4454	0·6214	0·84373	1·3953	1·2914	−0·3415
1·78	0·4413	0·6170	0·84142	1·3982	1·2993	−0·3478
1·79	0·4373	0·6127	0·83913	1·4011	1·3072	−0·3541
1·80	0·4333	0·6083	0·83685	1·4039	1·3153	−0·3603
1·81	0·4294	0·6040	0·83459	1·4067	1·3234	−0·3666
1·82	0·4255	0·5997	0·83233	1·4095	1·3317	−0·3729
1·83	0·4217	0·5955	0·83009	1·4122	1·3401	−0·3792
1·84	0·4179	0·5913	0·82787	1·4149	1·3487	−0·3856
1·85	0·4142	0·5871	0·82566	1·4175	1·3573	−0·3919
1·86	0·4105	0·5830	0·82346	1·4202	1·3661	−0·3983
1·87	0·4069	0·5789	0·82127	1·4228	1·3750	−0·4046
1·88	0·4033	0·5748	0·81910	1·4253	1·3840	−0·4109
1·89	0·3997	0·5707	0·81694	1·4279	1·3931	−0·4173
1·90	0·3962	0·5667	0·81480	1·4304	1·4023	−0·4236
1·91	0·3928	0·5627	0·81267	1·4328	1·4117	−0·4300
1·92	0·3893	0·5588	0·81055	1·4353	1·4212	−0·4363
1·93	0·3860	0·5549	0·80845	1·4377	1·4308	−0·4427
1·94	0·3826	0·5510	0·80636	1·4400	1·4406	−0·4491
1·95	0·3793	0·5471	0·80429	1·4424	1·4505	−0·4555
1·96	0·3761	0·5433	0·80223	1·4447	1·4605	−0·4618
1·97	0·3729	0·5395	0·80019	1·4470	1·4706	−0·4682
1·98	0·3697	0·5358	0·79816	1·4493	1·4809	−0·4745
1·99	0·3665	0·5320	0·79614	1·4515	1·4913	−0·4809
2·00	0·3634	0·5283	0·79414	1·4537	1·5018	−0·4873
2·01	0·3604	0·5247	0·79215	1·4559	1·5124	−0·4936
2·02	0·3573	0·5210	0·79018	1·4581	1·5232	−0·5000
2·03	0·3543	0·5174	0·78822	1·4602	1·5341	−0·5064
2·04	0·3514	0·5138	0·78627	1·4623	1·5452	−0·5127
2·05	0·3485	0·5103	0·78434	1·4644	1·5564	−0·5191
2·06	0·3456	0·5068	0·78242	1·4665	1·5677	−0·5254
2·07	0·3427	0·5033	0·78052	1·4685	1·5792	−0·5318
2·08	0·3399	0·4998	0·77863	1·4705	1·5908	−0·5381
2·09	0·3371	0·4964	0·77676	1·4725	1·6025	−0·5445
2·10	0·3343	0·4930	0·77489	1·4745	1·6144	−0·5508
2·11	0·3316	0·4896	0·77305	1·4764	1·6264	−0·5571
2·12	0·3289	0·4863	0·77121	1·4783	1·6386	−0·5635
2·13	0·3263	0·4829	0·76940	1·4802	1·6509	−0·5698
2·14	0·3236	0·4796	0·76759	1·4821	1·6633	−0·5762
2·15	0·3210	0·4764	0·76580	1·4839	1·6759	−0·5824
2·16	0·3185	0·4732	0·76402	1·4858	1·6886	−0·5888
2·17	0·3159	0·4699	0·76226	1·4876	1·7015	−0·5951
2·18	0·3134	0·4668	0·76050	1·4894	1·7146	−0·6014
2·19	0·3109	0·4636	0·75877	1·4912	1·7278	−0·6077
2·20	0·3085	0·4605	0·75704	1·4929	1·7411	−0·6140

TABLE II

FRICTIONLESS FLOW OF DRY AIR IN A CONSTANT AREA DUCT WITH HEAT TRANSFER

M	p/p^*	T/T^*	T_0/T_0^* $\left(1-\dfrac{q}{c_p T_0^*}\right)$	u/u^* (ρ^*/ρ)	P_0/P_0^*	$\Delta S/c_v$
2·21	0·3060	0·4574	0·75533	1·4946	1·7546	−0·6202
2·22	0·3036	0·4543	0·75363	1·4964	1·7682	−0·6265
2·23	0·3012	0·4513	0·75195	1·4980	1·7820	−0·6328
2·24	0·2989	0·4483	0·75028	1·4997	1·7960	−0·6391
2·25	0·2966	0·4453	0·74862	1·5014	1·8101	−0·6453
2·26	0·2943	0·4423	0·74698	1·5030	1·8243	−0·6515
2·27	0·2920	0·4393	0·74534	1·5046	1·8387	−0·6578
2·28	0·2898	0·4364	0·74372	1·5062	1·8533	−0·6640
2·29	0·2875	0·4335	0·74212	1·5078	1·8681	−0·6703
2·30	0·2853	0·4307	0·74052	1·5094	1·8830	−0·6765
2·31	0·2832	0·4278	0·73894	1·5109	1·8980	−0·6827
2·32	0·2810	0·4250	0·73737	1·5125	1·9132	−0·6889
2·33	0·2789	0·4222	0·73582	1·5140	1·9286	−0·6951
2·34	0·2768	0·4194	0·73427	1·5155	1·9442	−0·7013
2·35	0·2747	0·4167	0·73274	1·5170	1·9599	−0·7075
2·36	0·2726	0·4140	0·73122	1·5184	1·9758	−0·7136
2·37	0·2706	0·4113	0·72971	1·5199	1·9919	−0·7197
2·38	0·2686	0·4086	0·72822	1·5213	2·0081	−0·7259
2·39	0·2666	0·4059	0·72673	1·5227	2·0245	−0·7321
2·40	0·2646	0·4033	0·72526	1·5242	2·0411	−0·7382
2·41	0·2627	0·4007	0·72380	1·5255	2·0579	−0·7443
2·42	0·2607	0·3981	0·72235	1·5269	2·0748	−0·7504
2·43	0·2588	0·3955	0·72092	1·5283	2·0919	−0·7566
2·44	0·2569	0·3930	0·71949	1·5296	2·1092	−0·7627
2·45	0·2551	0·3905	0·71808	1·5310	2·1267	−0·7687
2·46	0·2532	0·3880	0·71667	1·5323	2·1444	−0·7748
2·47	0·2514	0·3855	0·71528	1·5336	2·1622	−0·7809
2·48	0·2496	0·3830	0·71390	1·5349	2·1802	−0·7869
2·49	0·2478	0·3806	0·71253	1·5362	2·1984	−0·7930
2·50	0·2460	0·3782	0·71117	1·5374	2·2169	−0·7990
2·51	0·2442	0·3758	0·70982	1·5387	2·2354	−0·8051
2·52	0·2425	0·3734	0·70849	1·5399	2·2542	−0·8111
2·53	0·2408	0·3711	0·70716	1·5411	2·2732	−0·8171
2·54	0·2391	0·3687	0·70584	1·5424	2·2924	−0·8231
2·55	0·2374	0·3664	0·70454	1·5436	2·3117	−0·8290
2·56	0·2357	0·3641	0·70324	1·5448	2·3313	−0·8350
2·57	0·2341	0·3618	0·70196	1·5459	2·3510	−0·8410
2·58	0·2324	0·3596	0·70068	1·5471	2·3710	−0·8469
2·59	0·2308	0·3573	0·69942	1·5483	2·3911	−0·8529
2·60	0·2292	0·3551	0·69817	1·5494	2·4115	−0·8588
2·61	0·2276	0·3529	0·69692	1·5505	2·4321	−0·8648
2·62	0·2260	0·3507	0·69569	1·5516	2·4528	−0·8707
2·63	0·2245	0·3486	0·69446	1·5528	2·4738	−0·8766
2·64	0·2229	0·3464	0·69325	1·5539	2·4950	−0·8824
2·65	0·2214	0·3443	0·69204	1·5549	2·5164	−0·8883

TABLE II

FRICTIONLESS FLOW OF DRY AIR IN A CONSTANT AREA DUCT WITH HEAT TRANSFER

M	p/p^*	T/T^*	T_0/T_0^* $\left(1-\dfrac{q}{C_p T_0^*}\right)$	u/u^* (ρ^*/ρ)	p_0/p_0^*	$\Delta S/c_v$
2·66	0·2199	0·3422	0·69085	1·5560	2·5380	−0·8942
2·67	0·2184	0·3401	0·68966	1·5571	2·5598	−0·9001
2·68	0·2169	0·3380	0·68849	1·5581	2·5818	−0·9059
2·69	0·2155	0·3360	0·68732	1·5592	2·6041	−0·9118
2·70	0·2140	0·3339	0·68616	1·5602	2·6266	−0·9176
2·71	0·2126	0·3319	0·68501	1·5612	2·6492	−0·9234
2·72	0·2112	0·3299	0·68387	1·5623	2·6721	−0·9292
2·73	0·2098	0·3279	0·68274	1·5633	2·6953	−0·9350
2·74	0·2084	0·3259	0·68162	1·5643	2·7186	−0·9408
2·75	0·2070	0·3240	0·68051	1·5652	2·7422	−0·9466
2·76	0·2056	0·3220	0·67940	1·5662	2·7660	−0·9523
2·77	0·2042	0·3201	0·67831	1·5672	2·7901	−0·9581
2·78	0·2029	0·3182	0·67722	1·5681	2·8143	−0·9638
2·79	0·2016	0·3163	0·67614	1·5691	2·8388	−0·9696
2·80	0·2003	0·3144	0·67507	1·5700	2·8636	−0·9753
2·81	0·1990	0·3125	0·67401	1·5710	2·8885	−0·9810
2·82	0·1977	0·3107	0·67296	1·5719	2·9138	−0·9866
2·83	0·1964	0·3089	0·67192	1·5728	2·9392	−0·9924
2·84	0·1951	0·3070	0·67088	1·5737	2·9649	−0·9980
2·85	0·1939	0·3052	0·66985	1·5746	2·9908	−1·004
2·86	0·1926	0·3035	0·66883	1·5755	3·0170	−1·009
2·87	0·1914	0·3017	0·66782	1·5764	3·0435	−1·015
2·88	0·1902	0·2999	0·66681	1·5772	3·0701	−1·021
2·89	0·1889	0·2982	0·66582	1·5781	3·0971	−1·026
2·90	0·1877	0·2964	0·66483	1·5789	3·1243	−1·032
2·91	0·1866	0·2947	0·66385	1·5798	3·1517	−1·037
2·92	0·1854	0·2930	0·66287	1·5806	3·1794	−1·043
2·93	0·1842	0·2913	0·66191	1·5815	3·2073	−1·049
2·94	0·1831	0·2896	0·66095	1·5823	3·2356	−1·054
2·95	0·1819	0·2880	0·66000	1·5831	3·2640	−1·060
2·96	0·1808	0·2863	0·65905	1·5839	3·2928	−1·065
2·97	0·1797	0·2847	0·65812	1·5847	3·3218	−1·071
2·98	0·1785	0·2831	0·65719	1·5855	3·3511	−1·076
2·99	0·1774	0·2815	0·65626	1·5863	3·3806	−1·082
3·00	0·1763	0·2799	0·65535	1·5871	3·4104	−1·087
3·01	0·1753	0·2783	0·65444	1·5878	3·4405	−1·093
3·02	0·1742	0·2767	0·65354	1·5886	3·4709	−1·098
3·03	0·1731	0·2751	0·65265	1·5894	3·5015	−1·104
3·04	0·1721	0·2736	0·65176	1·5901	3·5324	−1·109
3·05	0·1710	0·2721	0·65088	1·5909	3·5636	−1·115
3·06	0·1700	0·2705	0·65000	1·5916	3·5951	−1·120
3·07	0·1690	0·2690	0·64914	1·5923	3·6269	−1·125
3·08	0·1679	0·2675	0·64828	1·5931	3·6590	−1·131
3·09	0·1669	0·2660	0·64742	1·5938	3·6913	−1·136
3·10	0·1659	0·2646	0·64657	1·5945	3·7240	−1·142

TABLE II

FRICTIONLESS FLOW OF DRY AIR IN A CONSTANT AREA DUCT WITH HEAT TRANSFER

M	p/p^*	T/T^*	T_0/T_0^* $\left(1-\dfrac{q}{C_pT_0^*}\right)$	u/u^* (ρ^*/ρ)	p_0/p_0^*	$\Delta S/c_V$
3·11	0·1649	0·2631	0·64573	1·5952	3·7569	−1·147
3·12	0·1639	0·2616	0·64490	1·5959	3·7901	−1·152
3·13	0·1630	0·2602	0·64407	1·5966	3·8237	−1·158
3·14	0·1620	0·2588	0·64325	1·5973	3·8575	−1·163
3·15	0·1610	0·2573	0·64243	1·5980	3·8916	−1·168
3·16	0·1601	0·2559	0·64162	1·5986	3·9261	−1·174
3·17	0·1592	0·2545	0·64082	1·5993	3·9608	−1·179
3·18	0·1582	0·2531	0·64002	1·6000	3·9959	−1·184
3·19	0·1573	0·2518	0·63923	1·6006	4·0313	−1·190
3·20	0·1564	0·2504	0·63844	1·6013	4·0669	−1·195
3·21	0·1555	0·2491	0·63766	1·6019	4·1029	−1·200
3·22	0·1546	0·2477	0·63688	1·6026	4·1393	−1·205
3·23	0·1537	0·2464	0·63612	1·6032	4·1759	−1·211
3·24	0·1528	0·2450	0·63535	1·6039	4·2129	−1·216
3·25	0·1519	0·2437	0·63459	1·6045	4·2501	−1·221
3·26	0·1510	0·2424	0·63384	1·6051	4·2877	−1·226
3·27	0·1502	0·2411	0·63310	1·6057	4·3257	−1·232
3·28	0·1493	0·2398	0·63236	1·6063	4·3640	−1·237
3·29	0·1485	0·2386	0·63162	1·6069	4·4026	−1·242
3·30	0·1476	0·2373	0·63089	1·6075	4·4415	−1·247
3·31	0·1468	0·2360	0·63017	1·6081	4·4808	−1·252
3·32	0·1460	0·2348	0·62945	1·6087	4·5204	−1·257
3·33	0·1451	0·2336	0·62873	1·6093	4·5604	−1·263
3·34	0·1443	0·2323	0·62802	1·6099	4·6007	−1·268
3·35	0·1435	0·2311	0·62732	1·6105	4·6413	−1·273
3·36	0·1427	0·2299	0·62662	1·6110	4·6824	−1·278
3·37	0·1419	0·2287	0·62593	1·6116	4·7237	−1·283
3·38	0·1411	0·2275	0·62524	1·6122	4·7654	−1·288
3·39	0·1403	0·2263	0·62456	1·6127	4·8075	−1·293
3·40	0·1396	0·2251	0·62388	1·6133	4·8499	−1·298
3·41	0·1388	0·2240	0·62320	1·6138	4·8927	−1·303
3·42	0·1380	0·2228	0·62254	1·6144	4·9359	−1·308
3·43	0·1373	0·2217	0·62187	1·6149	4·9794	−1·313
3·44	0·1365	0·2205	0·62121	1·6155	5·0233	−1·318
3·45	0·1358	0·2194	0·62056	1·6160	5·0676	−1·323
3·46	0·1350	0·2183	0·61991	1·6165	5·1123	−1·328
3·47	0·1343	0·2172	0·61926	1·6170	5·1573	−1·333
3·48	0·1336	0·2161	0·61862	1·6176	5·2027	−1·338
3·49	0·1328	0·2150	0·61799	1·6181	5·2485	−1·343
3·50	0·1321	0·2139	0·61736	1·6186	5·2947	−1·348
3·51	0·1314	0·2128	0·61673	1·6191	5·3412	−1·353
3·52	0·1307	0·2117	0·61611	1·6196	5·3882	−1·358
3·53	0·1300	0·2106	0·61549	1·6201	5·4355	−1·363
3·54	0·1293	0·2096	0·61487	1·6206	5·4833	−1·368
3·55	0·1286	0·2085	0·61427	1·6211	5·5314	−1·373

TABLE II

FRICTIONLESS FLOW OF DRY AIR IN A CONSTANT AREA DUCT WITH HEAT TRANSFER

M	p/p^*	T/T^*	T_0/T_0^* $\left(1-\dfrac{q}{C_p T_0^*}\right)$	u/u^* (ρ^*/ρ)	p_0/p_0^*	$\Delta S/c_v$
3·56	0·1279	0·2075	0·61366	1·6216	5·5800	−1·378
3·57	0·1273	0·2064	0·61306	1·6220	5·6289	−1·383
3·58	0·1266	0·2054	0·61246	1·6225	5·6783	−1·388
3·59	0·1259	0·2044	0·61187	1·6230	5·7280	−1·393
3·60	0·1253	0·2034	0·61128	1·6235	5·7782	−1·397
3·61	0·1246	0·2024	0·61070	1·6239	5·8288	−1·402
3·62	0·1240	0·2014	0·61012	1·6244	5·8798	−1·407
3·63	0·1233	0·2004	0·60954	1·6249	5·9312	−1·412
3·64	0·1227	0·1994	0·60897	1·6253	5·9831	−1·417
3·65	0·1220	0·1984	0·60840	1·6258	6·0354	−1·422
3·66	0·1214	0·1974	0·60784	1·6262	6·0881	−1·426
3·67	0·1208	0·1965	0·60728	1·6267	6·1412	−1·431
3·68	0·1202	0·1955	0·60672	1·6271	6·1948	−1·436
3·69	0·1195	0·1945	0·60617	1·6276	6·2488	−1·441
3·70	0·1189	0·1936	0·60562	1·6280	6·3033	−1·446
3·71	0·1183	0·1927	0·60507	1·6284	6·3581	−1·450
3·72	0·1177	0·1917	0·60453	1·6289	6·4135	−1·455
3·73	0·1171	0·1908	0·60399	1·6293	6·4693	−1·460
3·74	0·1165	0·1899	0·60346	1·6297	6·5255	−1·465
3·75	0·1159	0·1890	0·60293	1·6301	6·5822	−1·469
3·76	0·1153	0·1881	0·60240	1·6306	6·6394	−1·474
3·77	0·1148	0·1872	0·60188	1·6310	6·6970	−1·479
3·78	0·1142	0·1863	0·60136	1·6314	6·7551	−1·483
3·79	0·1136	0·1854	0·60084	1·6318	6·8136	−1·488
3·80	0·1130	0·1845	0·60033	1·6322	6·8726	−1·493
3·81	0·1125	0·1836	0·59982	1·6326	6·9321	−1·497
3·82	0·1119	0·1827	0·59932	1·6330	6·9921	−1·502
3·83	0·1114	0·1819	0·59881	1·6334	7·0525	−1·507
3·84	0·1108	0·1810	0·59832	1·6338	7·1134	−1·511
3·85	0·1102	0·1802	0·59782	1·6342	7·1748	−1·516
3·86	0·1097	0·1793	0·59733	1·6346	7·2367	−1·521
3·87	0·1092	0·1785	0·59684	1·6350	7·2991	−1·525
3·88	0·1086	0·1776	0·59635	1·6353	7·3620	−1·530
3·89	0·1081	0·1768	0·59587	1·6357	7·4253	−1·534
3·90	0·1076	0·1760	0·59539	1·6361	7·4892	−1·539
3·91	0·1070	0·1752	0·59491	1·6365	7·5536	−1·543
3·92	0·1065	0·1744	0·59444	1·6368	7·6185	−1·548
3·93	0·1060	0·1735	0·59397	1·6372	7·6839	−1·553
3·94	0·1055	0·1727	0·59350	1·6376	7·7498	−1·557
3·95	0·1050	0·1719	0·59304	1·6379	7·8162	−1·562
3·96	0·1045	0·1712	0·59258	1·6383	7·8832	−1·566
3·97	0·1040	0·1704	0·59212	1·6387	7·9506	−1·571
3·98	0·1035	0·1696	0·59167	1·6390	8·0186	−1·575
3·99	0·1030	0·1688	0·59122	1·6394	8·0871	−1·580
4·00	0·1025	0·1680	0·59077	1·6397	8·1562	−1·584

TABLE III

ADIABATIC FLOW OF DRY AIR IN A CONSTANT AREA DUCT WITH SURFACE FRICTION

M	p/p^*	p_0/p_0^*	ρ/ρ^* (u^*/u)	T/T^*	$\Delta S/c_v$	$\dfrac{C_f 4\lvert x-x^*\rvert}{d}$
0·01	109·612	57·855	91·231	1·2015	−1·635	7119·14
0·02	54·804	28·933	45·617	1·2014	−1·356	1774·64
0·03	36·534	19·294	30·413	1·2013	−1·193	785·392
0·04	27·399	14·477	22·811	1·2011	−1·077	439·405
0·05	21·917	11·588	18·251	1·2009	−0·9873	279·416
0·06	18·262	9·663	15·211	1·2006	−0·9141	192·614
0·07	15·651	8·289	13·039	1·2003	−0·8523	140·350
0·08	13·693	7·259	11·411	1·2000	−0·7988	106·486
0·09	12·169	6·459	10·145	1·1995	−0·7518	83·3137
0·10	10·950	5·820	9·132	1·1991	−0·7098	66·7748
0·11	9·953	5·298	8·304	1·1986	−0·6719	54·5675
0·12	9·121	4·863	7·614	1·1980	−0·6374	45·3076
0·13	8·417	4·495	7·030	1·1974	−0·6057	38·1222
0·14	7·814	4·181	6·529	1·1968	−0·5765	32·4388
0·15	7·291	3·909	6·096	1·1961	−0·5494	27·8694
0·16	6·833	3·672	5·717	1·1953	−0·5241	24·1434
0·17	6·429	3·462	5·382	1·1945	−0·5005	21·0675
0·18	6·070	3·277	5·085	1·1937	−0·4783	18·5006
0·19	5·748	3·111	4·819	1·1928	−0·4574	16·3378
0·20	5·459	2·963	4·580	1·1919	−0·4377	14·5000
0·21	5·197	2·828	4·364	1·1909	−0·4190	12·9262
0·22	4·958	2·707	4·167	1·1899	−0·4013	11·5692
0·23	4·741	2·596	3·988	1·1888	−0·3844	10·3919
0·24	4·541	2·495	3·823	1·1877	−0·3685	9·36454
0·25	4·357	2·402	3·672	1·1866	−0·3532	8·46348
0·26	4·187	2·317	3·533	1·1854	−0·3385	7·66941
0·27	4·030	2·238	3·404	1·1841	−0·3246	6·96659
0·28	3·884	2·165	3·284	1·1828	−0·3113	6·34205
0·29	3·748	2·097	3·172	1·1815	−0·2985	5·78501
0·30	3·621	2·035	3·068	1·1801	−0·2862	5·28648
0·31	3·502	1·976	2·971	1·1787	−0·2745	4·83891
0·32	3·391	1·921	2·880	1·1772	−0·2631	4·43591
0·33	3·286	1·870	2·795	1·1757	−0·2523	4·07206
0·34	3·187	1·822	2·714	1·1742	−0·2419	3·74272
0·35	3·094	1·778	2·639	1·1726	−0·2318	3·44391
0·36	3·006	1·735	2·567	1·1709	−0·2222	3·17221
0·37	2·922	1·696	2·499	1·1692	−0·2128	2·92465
0·38	2·843	1·658	2·435	1·1675	−0·2038	2·69865
0·39	2·768	1·623	2·375	1·1658	−0·1952	2·49196
0·40	2·697	1·590	2·317	1·1640	−0·1868	2·30263
0·41	2·629	1·558	2·262	1·1621	−0·1788	2·12891
0·42	2·565	1·529	2·210	1·1603	−0·1710	1·96930
0·43	2·503	1·500	2·161	1·1583	−0·1635	1·82244
0·44	2·444	1·474	2·113	1·1564	−0·1563	1·68714

27

TABLE III

DIABATIC FLOW OF DRY AIR IN A CONSTANT AREA DUCT WITH SURFACE FRICTION

M	p/p^*	p_0/p_0^*	ρ/ρ^* (u^*/u)	T/T^*	$\Delta S/c_v$	$C_f\, 4\, \dfrac{\|x-x^*\|}{d}$
0·45	2·388	1·448	2·068	1·1544	−0·1493	1·56234
0·46	2·334	1·424	2·025	1·1524	−0·1426	1·44711
0·47	2·282	1·402	1·984	1·1503	−0·1360	1·34059
0·48	2·232	1·380	1·944	1·1482	−0·1297	1·24204
0·49	2·185	1·359	1·906	1·1461	−0·1237	1·15078
0·50	2·139	1·340	1·870	1·1439	−0·1178	1·06620
0·51	2·095	1·321	1·835	1·1417	−0·1122	0·98775
0·52	2·053	1·303	1·802	1·1394	−0·1067	0·91494
0·53	2·012	1·286	1·769	1·1371	−0·1015	0·84731
0·54	1·973	1·270	1·738	1·1348	−0·0964	0·78448
0·55	1·935	1·255	1·709	1·1325	−0·0915	0·72605
0·56	1·898	1·240	1·680	1·1301	−0·0867	0·67171
0·57	1·863	1·226	1·652	1·1277	−0·0822	0·62115
0·58	1·829	1·213	1·625	1·1252	−0·0778	0·57408
0·59	1·796	1·200	1·600	1·1227	−0·0735	0·53025
0·60	1·764	1·188	1·575	1·1202	−0·0694	0·48944
0·61	1·733	1·177	1·551	1·1177	−0·0655	0·45142
0·62	1·703	1·166	1·527	1·1151	−0·0617	0·41601
0·63	1·674	1·155	1·505	1·1125	−0·0581	0·38302
0·64	1·646	1·145	1·483	1·1099	−0·0546	0·35228
0·65	1·619	1·136	1·462	1·1072	−0·0512	0·32365
0·66	1·592	1·126	1·442	1·1045	−0·0480	0·29699
0·67	1·567	1·118	1·422	1·1018	−0·0449	0·27216
0·68	1·542	1·110	1·403	1·0991	−0·0419	0·24904
0·69	1·517	1·102	1·384	1·0963	−0·0391	0·22753
0·70	1·494	1·094	1·366	1·0935	−0·0363	0·20752
0·71	1·471	1·087	1·349	1·0907	−0·0337	0·18892
0·72	1·449	1·080	1·332	1·0879	−0·0312	0·17163
0·73	1·427	1·074	1·315	1·0850	−0·0288	0·15559
0·74	1·406	1·068	1·299	1·0821	−0·0265	0·14070
0·75	1·385	1·062	1·283	1·0792	−0·0244	0·12690
0·76	1·365	1·057	1·268	1·0762	−0·0223	0·11412
0·77	1·345	1·052	1·254	1·0733	−0·0203	0·10230
0·78	1·326	1·047	1·239	1·0703	−0·0185	0·09139
0·79	1·308	1·042	1·225	1·0673	−0·0167	0·08133
0·80	1·290	1·038	1·212	1·0643	−0·0151	0·07207
0·81	1·272	1·034	1·198	1·0612	−0·0135	0·06356
0·82	1·254	1·030	1·186	1·0581	−0·0121	0·05576
0·83	1·238	1·027	1·173	1·0550	−0·0107	0·04862
0·84	1·221	1·024	1·161	1·0519	−0·0094	0·04212
0·85	1·205	1·021	1·149	1·0488	−0·0082	0·03621
0·86	1·189	1·018	1·137	1·0457	−0·0071	0·03087
0·87	1·174	1·015	1·126	1·0425	−0·0061	0·02605
0·88	1·158	1·013	1·115	1·0393	−0·0052	0·02172
0·89	1·144	1·011	1·104	1·0361	−0·0043	0·01787

TABLE III

ADIABATIC FLOW OF DRY AIR IN A CONSTANT AREA DUCT WITH SURFACE FRICTION

M	p/p^*	p_0/p_0^*	ρ/ρ^* (u^*/u)	T/T^*	$\Delta S/c_v$	$\dfrac{C_f 4 \lvert x-x^* \rvert}{d}$
0·90	1·129	1·009	1·093	1·0329	−0·0036	0·01447
0·91	1·115	1·007	1·083	1·0297	−0·0029	0·01148
0·92	1·101	1·006	1·073	1·0264	−0·0023	0·00888
0·93	1·088	1·004	1·063	1·0232	−0·0017	0·00667
0·94	1·074	1·003	1·053	1·0199	−0·0012	0·00480
0·95	1·061	1·002	1·044	1·0166	−0·0009	0·00327
0·96	1·049	1·001	1·035	1·0133	−0·0005	0·00205
0·97	1·036	1·001	1·026	1·0100	−0·0003	0·00113
0·98	1·024	1·000	1·017	1·0067	−0·0001	0·00049
0·99	1·012	1·000	1·008	1·0033	−0·0001	0·00012
1·00	1·0000	1·0000	1·0000	1·0000	0·0000	0·00000
1·01	0·9884	1·0001	0·9918	0·9966	−0·0001	0·00012
1·02	0·9771	1·0003	0·9837	0·9933	−0·0001	0·00046
1·03	0·9660	1·0007	0·9758	0·9899	−0·0003	0·00101
1·04	0·9550	1·0013	0·9681	0·9865	−0·0005	0·00176
1·05	0·9443	1·0020	0·9605	0·9831	−0·0008	0·00270
1·06	0·9338	1·0029	0·9531	0·9797	−0·0012	0·00382
1·07	0·9234	1·0039	0·9459	0·9763	−0·0016	0·00511
1·08	0·9153	1·0051	0·9388	0·9729	−0·0020	0·00656
1·09	0·9033	1·0064	0·9318	0·9694	−0·0026	0·00816
1·10	0·8935	1·0079	0·9250	0·9660	−0·0031	0·00990
1·11	0·8839	1·0095	0·9183	0·9625	−0·0038	0·01177
1·12	0·8744	1·0113	0·9117	0·9591	−0·0045	0·01377
1·13	0·8651	1·0132	0·9053	0·9556	−0·0053	0·01589
1·14	0·8560	1·0152	0·8990	0·9522	−0·0061	0·01812
1·15	0·8470	1·0174	0·8928	0·9487	−0·0069	0·02046
1·16	0·8381	1·0198	0·8867	0·9452	−0·0079	0·02289
1·17	0·8294	1·0222	0·8807	0·9417	−0·0089	0·02542
1·18	0·8209	1·0248	0·8749	0·9383	−0·0099	0·02804
1·19	0·8125	1·0275	0·8692	0·9348	−0·0110	0·03074
1·20	0·8042	1·0304	0·8635	0·9313	−0·0121	0·03352
1·21	0·7960	1·0334	0·8580	0·9278	−0·0132	0·03636
1·22	0·7880	1·0365	0·8526	0·9243	−0·0145	0·03928
1·23	0·7801	1·0398	0·8473	0·9208	−0·0157	0·04226
1·24	0·7724	1·0432	0·8420	0·9173	−0·0170	0·04530
1·25	0·7647	1·0467	0·8369	0·9138	−0·0184	0·04840
1·26	0·7572	1·0503	0·8318	0·9103	−0·0198	0·05155
1·27	0·7498	1·0541	0·8269	0·9068	−0·0212	0·05474
1·28	0·7425	1·0580	0·8220	0·9033	−0·0227	0·05798
1·29	0·7353	1·0620	0·8172	0·8998	−0·0243	0·06127
1·30	0·7282	1·0662	0·8125	0·8963	−0·0258	0·06459

TABLE III

ADIABATIC FLOW OF DRY AIR IN A CONSTANT AREA DUCT WITH SURFACE FRICTION

M	p/p^*	p_0/p_0^*	ρ/ρ^* (u^*/u)	T/T^*	$\Delta S/c_v$	$\dfrac{C_f 4 \lvert x-x^* \rvert}{d}$
1·31	0·7213	1·0705	0·8079	0·8928	−0·0275	0·06794
1·32	0·7144	1·0749	0·8034	0·8893	−0·0291	0·07133
1·33	0·7076	1·0794	0·7989	0·8858	−0·0308	0·07475
1·34	0·7010	1·0841	0·7945	0·8823	−0·0325	0·07820
1·35	0·6944	1·0889	0·7902	0·8788	−0·0343	0·08168
1·36	0·6879	1·0938	0·7859	0·8753	−0·0362	0·08517
1·37	0·6815	1·0988	0·7818	0·8718	−0·0380	0·08869
1·38	0·6752	1·1040	0·7777	0·8683	−0·0399	0·09223
1·39	0·6690	1·1093	0·7736	0·8648	−0·0418	0·09578
1·40	0·6629	1·1147	0·7696	0·8613	−0·0438	0·09935
1·41	0·6569	1·1202	0·7657	0·8578	−0·0458	0·10293
1·42	0·6509	1·1259	0·7619	0·8544	−0·0478	0·10652
1·43	0·6451	1·1317	0·7581	0·8509	−0·0499	0·11013
1·44	0·6393	1·1376	0·7544	0·8474	−0·0520	0·11374
1·45	0·6336	1·1437	0·7507	0·8440	−0·0541	0·11736
1·46	0·6279	1·1498	0·7471	0·8405	−0·0563	0·12098
1·47	0·6224	1·1561	0·7436	0·8370	−0·0585	0·12461
1·48	0·6169	1·1626	0·7401	0·8336	−0·0607	0·12824
1·49	0·6115	1·1691	0·7366	0·8301	−0·0630	0·13187
1·50	0·6062	1·1758	0·7332	0·8267	−0·0652	0·13550
1·51	0·6009	1·1826	0·7299	0·8233	−0·0676	0·13914
1·52	0·5957	1·1895	0·7266	0·8198	−0·0700	0·14277
1·53	0·5906	1·1966	0·7234	0·8164	−0·0723	0·14639
1·54	0·5855	1·2038	0·7202	0·8130	−0·0748	0·15002
1·55	0·5805	1·2111	0·7170	0·8096	−0·0772	0·15364
1·56	0·5756	1·2186	0·7139	0·8062	−0·0797	0·15725
1·57	0·5707	1·2261	0·7109	0·8028	−0·0822	0·16086
1·58	0·5659	1·2338	0·7079	0·7994	−0·0847	0·16446
1·59	0·5611	1·2417	0·7049	0·7960	−0·0872	0·16806
1·60	0·5564	1·2497	0·7020	0·7926	−0·0898	0·17164
1·61	0·5518	1·2578	0·6991	0·7893	−0·0924	0·17522
1·62	0·5472	1·2660	0·6963	0·7859	−0·0950	0·17879
1·63	0·5427	1·2744	0·6935	0·7825	−0·0977	0·18235
1·64	0·5382	1·2829	0·6908	0·7792	−0·1004	0·18589
1·65	0·5338	1·2915	0·6881	0·7759	−0·1031	0·18943
1·66	0·5295	1·3003	0·6854	0·7725	−0·1058	0·19295
1·67	0·5252	1·3092	0·6827	0·7692	−0·1086	0·19646
1·68	0·5209	1·3182	0·6801	0·7659	−0·1114	0·19996
1·69	0·5167	1·3274	0·6776	0·7626	−0·1142	0·20345
1·70	0·5126	1·3367	0·6751	0·7593	−0·1170	0·20692
1·71	0·5085	1·3462	0·6726	0·7560	−0·1198	0·21038
1·72	0·5044	1·3558	0·6701	0·7528	−0·1227	0·21382
1·73	0·5004	1·3656	0·6677	0·7495	−0·1255	0·21725
1·74	0·4965	1·3754	0·6653	0·7462	−0·1285	0·22067
1·75	0·4926	1·3855	0·6629	0·7430	−0·1314	0·22407

TABLE III

ADIABATIC FLOW OF DRY AIR IN A CONSTANT AREA DUCT WITH SURFACE FRICTION

M	p/p^*	p_0/p_0^*	ρ/ρ^* (u^*/u)	T/T^*	$\Delta S/c_V$	$\dfrac{C_f 4\,|x-x^*|}{d}$
1·76	0·4887	1·3956	0·6606	0·7398	−0·1343	0·22745
1·77	0·4849	1·4059	0·6583	0·7365	−0·1373	0·23082
1·78	0·4811	1·4164	0·6560	0·7333	−0·1403	0·23418
1·79	0·4774	1·4270	0·6538	0·7301	−0·1433	0·23751
1·80	0·4737	1·4378	0·6516	0·7269	−0·1463	0·24083
1·81	0·4700	1·4487	0·6494	0·7237	−0·1494	0·24414
1·82	0·4664	1·4597	0·6473	0·7206	−0·1524	0·24743
1·83	0·4628	1·4709	0·6452	0·7174	−0·1555	0·25070
1·84	0·4593	1·4823	0·6431	0·7142	−0·1586	0·25395
1·85	0·4558	1·4938	0·6410	0·7111	−0·1617	0·25719
1·86	0·4524	1·5054	0·6390	0·7080	−0·1648	0·26041
1·87	0·4490	1·5172	0·6370	0·7048	−0·1680	0·26361
1·88	0·4456	1·5292	0·6350	0·7017	−0·1711	0·26679
1·89	0·4422	1·5413	0·6330	0·6986	−0·1743	0·26996
1·90	0·4389	1·5536	0·6311	0·6955	−0·1776	0·27311
1·91	0·4357	1·5660	0·6292	0·6925	−0·1808	0·27624
1·92	0·4324	1·5786	0·6273	0·6894	−0·1840	0·27935
1·93	0·4293	1·5914	0·6254	0·6863	−0·1872	0·28245
1·94	0·4261	1·6043	0·6236	0·6833	−0·1905	0·28553
1·95	0·4230	1·6173	0·6218	0·6803	−0·1938	0·28859
1·96	0·4199	1·6306	0·6200	0·6773	−0·1970	0·29163
1·97	0·4168	1·6440	0·6182	0·6742	−0·2003	0·29465
1·98	0·4138	1·6576	0·6164	0·6712	−0·2037	0·29766
1·99	0·4108	1·6713	0·6147	0·6683	−0·2070	0·30064
2·00	0·4078	1·6852	0·6130	0·6653	−0·2103	0·30361
2·01	0·4049	1·6993	0·6113	0·6623	−0·2137	0·30656
2·02	0·4020	1·7136	0·6097	0·6594	−0·2170	0·30950
2·03	0·3991	1·7280	0·6080	0·6564	−0·2204	0·31241
2·04	0·3963	1·7426	0·6064	0·6535	−0·2238	0·31531
2·05	0·3935	1·7574	0·6048	0·6506	−0·2272	0·31819
2·06	0·3907	1·7723	0·6032	0·6477	−0·2306	0·32105
2·07	0·3879	1·7875	0·6016	0·6448	−0·2341	0·32389
2·08	0·3852	1·8028	0·6001	0·6419	−0·2375	0·32671
2·09	0·3825	1·8183	0·5985	0·6390	−0·2409	0·32952
2·10	0·3798	1·8339	0·5970	0·6362	−0·2444	0·33231
2·11	0·3772	1·8498	0·5955	0·6333	−0·2479	0·33508
2·12	0·3745	1·8658	0·5940	0·6305	−0·2514	0·33783
2·13	0·3720	1·8821	0·5926	0·6277	−0·2549	0·34056
2·14	0·3694	1·8985	0·5911	0·6249	−0·2584	0·34328
2·15	0·3668	1·9151	0·5897	0·6221	−0·2619	0·34598
2·16	0·3643	1·9319	0·5883	0·6193	−0·2654	0·34866
2·17	0·3618	1·9489	0·5869	0·6165	−0·2689	0·35133
2·18	0·3594	1·9661	0·5855	0·6138	−0·2724	0·35397
2·19	0·3569	1·9835	0·5842	0·6110	−0·2760	0·35660
2·20	0·3545	2·0011	0·5828	0·6083	−0·2795	0·35921

TABLE III

ADIABATIC FLOW OF DRY AIR IN A CONSTANT AREA DUCT WITH SURFACE FRICTION

M	p/p^*	p_0/p_0^*	ρ/ρ^* (u^*/u)	T/T^*	$\Delta S/c_V$	$\dfrac{C_f 4\|x-x^*\|}{d}$
2·21	0·3521	2·0188	0·5815	0·6056	−0·2831	0·36181
2·22	0·3497	2·0368	0·5802	0·6028	−0·2867	0·36438
2·23	0·3474	2·0550	0·5789	0·6001	−0·2903	0·36694
2·24	0·3451	2·0734	0·5776	0·5975	−0·2939	0·36948
2·25	0·3428	2·0920	0·5763	0·5948	−0·2974	0·37201
2·26	0·3405	2·1108	0·5750	0·5921	−0·3010	0·37452
2·27	0·3382	2·1298	0·5738	0·5895	−0·3047	0·37701
2·28	0·3360	2·1490	0·5725	0·5868	−0·3083	0·37948
2·29	0·3338	2·1685	0·5713	0·5842	−0·3120	0·38194
2·30	0·3316	2·1881	0·5701	0·5816	−0·3156	0·38438
2·31	0·3294	2·2080	0·5689	0·5790	−0·3192	0·38681
2·32	0·3272	2·2281	0·5677	0·5764	−0·3229	0·38922
2·33	0·3251	2·2484	0·5666	0·5738	−0·3265	0·39161
2·34	0·3230	2·2689	0·5654	0·5712	−0·3301	0·39398
2·35	0·3209	2·2896	0·5643	0·5687	−0·3338	0·39634
2·36	0·3188	2·3106	0·5632	0·5661	−0·3375	0·39869
2·37	0·3168	2·3318	0·5620	0·5636	−0·3412	0·40101
2·38	0·3147	2·3532	0·5609	0·5611	−0·3449	0·40332
2·39	0·3127	2·3749	0·5598	0·5586	−0·3486	0·40562
2·40	0·3107	2·3967	0·5588	0·5561	−0·3522	0·40790
2·41	0·3087	2·4189	0·5577	0·5536	−0·3560	0·41016
2·42	0·3068	2·4412	0·5566	0·5511	−0·3597	0·41241
2·43	0·3048	2·4638	0·5556	0·5487	−0·3634	0·41464
2·44	0·3029	2·4866	0·5545	0·5462	−0·3671	0·41686
2·45	0·3010	2·5097	0·5535	0·5438	−0·3708	0·41906
2·46	0·2991	2·5330	0·5525	0·5414	−0·3746	0·42125
2·47	0·2972	2·5566	0·5515	0·5390	−0·3783	0·42342
2·48	0·2954	2·5804	0·5505	0·5366	−0·3820	0·42558
2·49	0·2935	2·6044	0·5495	0·5342	−0·3858	0·42772
2·50	0·2917	2·6288	0·5485	0·5318	−0·3895	0·42985
2·51	0·2899	2·6533	0·5476	0·5294	−0·3932	0·43196
2·52	0·2881	2·6781	0·5466	0·5271	−0·3970	0·43406
2·53	0·2863	2·7032	0·5457	0·5247	−0·4008	0·43614
2·54	0·2846	2·7285	0·5447	0·5224	−0·4045	0·43821
2·55	0·2828	2·7541	0·5438	0·5201	−0·4083	0·44027
2·56	0·2811	2·7800	0·5429	0·5178	−0·4121	0·44231
2·57	0·2794	2·8061	0·5420	0·5155	−0·4158	0·44433
2·58	0·2777	2·8325	0·5411	0·5132	−0·4196	0·44635
2·59	0·2760	2·8592	0·5402	0·5109	−0·4234	0·44834
2·60	0·2743	2·8861	0·5393	0·5086	−0·4271	0·45033
2·61	0·2727	2·9133	0·5384	0·5064	−0·4309	0·45230
2·62	0·2710	2·9408	0·5375	0·5042	−0·4347	0·45426
2·63	0·2694	2·9685	0·5367	0·5019	−0·4385	0·45620
2·64	0·2678	2·9966	0·5358	0·4997	−0·4423	0·45813
2·65	0·2662	3·0249	0·5350	0·4975	−0·4461	0·46005

32

TABLE III

ADIABATIC FLOW OF DRY AIR IN A CONSTANT AREA DUCT WITH SURFACE FRICTION

M	p/p^*	$\dot{p}_0/p_0^*$	ρ/ρ^* (u^*/u)	T/T^*	$\Delta S/c_v$	$C_f 4 \left\lvert \dfrac{x-x^*}{d} \right.$
2·66	0·2646	3·0535	0·5342	0·4953	−0·4499	0·46195
2·67	0·2630	3·0824	0·5333	0·4931	−0·4536	0·46384
2·68	0·2614	3·1116	0·5325	0·4910	−0·4574	0·46572
2·69	0·2599	3·1411	0·5317	0·4888	−0·4613	0·46758
2·70	0·2584	3·1708	0·5309	0·4866	−0·4651	0·46944
2·71	0·2569	3·2009	0·5301	0·4845	−0·4689	0·47127
2·72	0·2553	3·2313	0·5293	0·4824	−0·4727	0·47310
2·73	0·2538	3·2619	0·5286	0·4803	−0·4765	0·47491
2·74	0·2524	3·2929	0·5278	0·4782	−0·4803	0·47671
2·75	0·2509	3·3242	0·5270	0·4761	−0·4841	0·47850
2·76	0·2494	3·3558	0·5263	0·4740	−0·4879	0·48028
2·77	0·2480	3·3877	0·5255	0·4719	−0·4917	0·48204
2·78	0·2466	3·4199	0·5248	0·4698	−0·4955	0·48380
2·79	0·2451	3·4524	0·5241	0·4678	−0·4994	0·48553
2·80	0·2437	3·4852	0·5233	0·4657	−0·5031	0·48726
2·81	0·2423	3·5184	0·5226	0·4637	−0·5070	0·48898
2·82	0·2409	3·5519	0·5219	0·4617	−0·5108	0·49068
2·83	0·2396	3·5857	0·5212	0·4597	−0·5146	0·49237
2·84	0·2382	3·6198	0·5205	0·4577	−0·5184	0·49406
2·85	0·2369	3·6543	0·5198	0·4557	−0·5223	0·49572
2·86	0·2355	3·6891	0·5191	0·4537	−0·5261	0·49738
2·87	0·2342	3·7243	0·5184	0·4517	−0·5299	0·49903
2·88	0·2329	3·7597	0·5177	0·4498	−0·5337	0·50066
2·89	0·2316	3·7955	0·5171	0·4478	−0·5376	0·50229
2·90	0·2303	3·8317	0·5164	0·4459	−0·5413	0·50390
2·91	0·2290	3·8682	0·5157	0·4440	−0·5452	0·50550
2·92	0·2277	3·9051	0·5151	0·4420	−0·5490	0·50709
2·93	0·2264	3·9423	0·5144	0·4401	−0·5528	0·50867
2·94	0·2252	3·9798	0·5138	0·4382	−0·5566	0·51024
2·95	0·2239	4·0178	0·5132	0·4363	−0·5605	0·51180
2·96	0·2227	4·0560	0·5125	0·4345	−0·5643	0·51335
2·97	0·2215	4·0947	0·5119	0·4326	−0·5681	0·51489
2·98	0·2202	4·1337	0·5113	0·4307	−0·5719	0·51641
2·99	0·2190	4·1731	0·5107	0·4289	−0·5757	0·51793
3·00	0·2178	4·2128	0·5101	0·4270	−0·5796	0·51944
3·01	0·2166	4·2529	0·5095	0·4252	−0·5834	0·52093
3·02	0·2155	4·2934	0·5089	0·4234	−0·5872	0·52242
3·03	0·2143	4·3343	0·5083	0·4216	−0·5910	0·52390
3·04	0·2131	4·3755	0·5077	0·4198	−0·5949	0·52536
3·05	0·2120	4·4172	0·5071	0·4180	−0·5987	0·52682
3·06	0·2108	4·4592	0·5066	0·4162	−0·6024	0·52826
3·07	0·2097	4·5016	0·5060	0·4144	−0·6063	0·52970
3·08	0·2086	4·5444	0·5054	0·4127	−0·6101	0·53113
3·09	0·2075	4·5876	0·5049	0·4109	−0·6139	0·53254
3·10	0·2063	4·6312	0·5043	0·4092	−0·6177	0·53395

TABLE III

ADIABATIC FLOW OF DRY AIR IN A CONSTANT AREA DUCT WITH SURFACE FRICTION

M	p/p^*	p_0/p_0^*	ρ/ρ^* (u^*/u)	T/T^*	$\Delta S/c_v$	$\dfrac{C_f\,4}{d}\Big\vert x - x^*$
3·11	0·2052	4·6752	0·5037	0·4074	−0·6215	0·53535
3·12	0·2042	4·7197	0·5032	0·4057	−0·6254	0·53674
3·13	0·2031	4·7645	0·5027	0·4040	−0·6292	0·53812
3·14	0·2020	4·8097	0·5021	0·4023	−0·6329	0·53949
3·15	0·2009	4·8553	0·5016	0·4006	−0·6368	0·54085
3·16	0·1999	4·9014	0·5011	0·3989	−0·6406	0·54220
3·17	0·1988	4·9479	0·5005	0·3972	−0·6444	0·54355
3·18	0·1978	4·9948	0·5000	0·3955	−0·6482	0·54488
3·19	0·1967	5·0421	0·4995	0·3939	−0·6520	0·54621
3·20	0·1957	5·0899	0·4990	0·3922	−0·6558	0·54752
3·21	0·1947	5·1381	0·4985	0·3906	−0·6596	0·54883
3·22	0·1937	5·1867	0·4980	0·3889	−0·6634	0·55013
3·23	0·1927	5·2358	0·4975	0·3873	−0·6672	0·55142
3·24	0·1917	5·2853	0·4970	0·3857	−0·6710	0·55270
3·25	0·1907	5·3352	0·4965	0·3841	−0·6748	0·55398
3·26	0·1897	5·3857	0·4960	0·3825	−0·6785	0·55524
3·27	0·1887	5·4365	0·4955	0·3809	−0·6823	0·55650
3·28	0·1878	5·4878	0·4950	0·3793	−0·6861	0·55775
3·29	0·1868	5·5396	0·4946	0·3777	−0·6899	0·55899
3·30	0·1858	5·5918	0·4941	0·3761	−0·6937	0·56022
3·31	0·1849	5·6445	0·4936	0·3746	−0·6975	0·56144
3·32	0·1840	5·6977	0·4932	0·3730	−0·7012	0·56266
3·33	0·1830	5·7513	0·4927	0·3715	−0·7050	0·56387
3·34	0·1821	5·8054	0·4923	0·3699	−0·7088	0·56507
3·35	0·1812	5·8600	0·4918	0·3684	−0·7125	0·56626
3·36	0·1803	5·9151	0·4914	0·3669	−0·7163	0·56744
3·37	0·1794	5·9706	0·4909	0·3654	−0·7201	0·56862
3·38	0·1785	6·0267	0·4905	0·3639	−0·7238	0·56979
3·39	0·1776	6·0832	0·4900	0·3624	−0·7276	0·57095
3·40	0·1767	6·1403	0·4896	0·3609	−0·7314	0·57211
3·41	0·1758	6·1978	0·4892	0·3594	−0·7352	0·57325
3·42	0·1749	6·2558	0·4887	0·3579	−0·7389	0·57439
3·43	0·1741	6·3144	0·4883	0·3565	−0·7427	0·57552
3·44	0·1732	6·3734	0·4879	0·3550	−0·7464	0·57665
3·45	0·1723	6·4330	0·4875	0·3536	−0·7501	0·57777
3·46	0·1715	6·4931	0·4871	0·3521	−0·7539	0·57888
3·47	0·1707	6·5537	0·4867	0·3507	−0·7577	0·57998
3·48	0·1698	6·6148	0·4862	0·3492	−0·7614	0·58107
3·49	0·1690	6·6764	0·4858	0·3478	−0·7651	0·58216
3·50	0·1682	6·7386	0·4854	0·3464	−0·7689	0·58325
3·51	0·1673	6·8013	0·4850	0·3450	−0·7726	0·58432
3·52	0·1665	6·8646	0·4846	0·3436	−0·7763	0·58539
3·53	0·1657	6·9284	0·4843	0·3422	−0·7801	0·58645
3·54	0·1649	6·9927	0·4839	0·3408	−0·7838	0·58750
3·55	0·1641	7·0576	0·4835	0·3395	−0·7875	0·58855

34

TABLE III

ADIABATIC FLOW OF DRY AIR IN A CONSTANT AREA DUCT WITH SURFACE FRICTION

M	p/p^*	p_0/p_0^*	ρ/ρ^* (u^*/u)	T/T^*	$\Delta S/c_v$	$C_f 4 \dfrac{\lvert x - x^* \rvert}{d}$
3·56	0·1633	7·1231	0·4831	0·3381	−0·7913	0·58959
3·57	0·1625	7·1891	0·4827	0·3367	−0·7949	0·59063
3·58	0·1618	7·2556	0·4823	0·3354	−0·7987	0·59166
3·59	0·1610	7·3228	0·4820	0·3340	−0·8024	0·59268
3·60	0·1602	7·3905	0·4816	0·3327	−0·8061	0·59369
3·61	0·1595	7·4587	0·4812	0·3314	−0·8098	0·59470
3·62	0·1587	7·5276	0·4809	0·3300	−0·8135	0·59570
3·63	0·1579	7·5970	0·4805	0·3287	−0·8172	0·59670
3·64	0·1572	7·6670	0·4801	0·3274	−0·8209	0·59769
3·65	0·1565	7·7376	0·4798	0·3261	−0·8246	0·59867
3·66	0·1557	7·8088	0·4794	0·3248	−0·8283	0·59965
3·67	0·1550	7·8806	0·4791	0·3235	−0·8320	0·60062
3·68	0·1543	7·9530	0·4787	0·3222	−0·8356	0·60159
3·69	0·1535	8·0260	0·4784	0·3209	−0·8393	0·60255
3·70	0·1528	8·0996	0·4780	0·3197	−0·8430	0·60350
3·71	0·1521	8·1738	0·4777	0·3184	−0·8467	0·60445
3·72	0·1514	8·2487	0·4773	0·3171	−0·8504	0·60539
3·73	0·1507	8·3241	0·4770	0·3159	−0·8540	0·60632
3·74	0·1500	8·4002	0·4767	0·3147	−0·8577	0·60725
3·75	0·1493	8·4769	0·4763	0·3134	−0·8613	0·60818
3·76	0·1486	8·5543	0·4760	0·3122	−0·8650	0·60909
3·77	0·1479	8·6322	0·4757	0·3110	−0·8687	0·61001
3·78	0·1472	8·7109	0·4753	0·3097	−0·8723	0·61091
3·79	0·1466	8·7901	0·4750	0·3085	−0·8760	0·61182
3·80	0·1459	8·8700	0·4747	0·3073	−0·8796	0·61271
3·81	0·1452	8·9506	0·4744	0·3061	−0·8833	0·61360
3·82	0·1446	9·0318	0·4741	0·3049	−0·8869	0·61449
3·83	0·1439	9·1137	0·4738	0·3037	−0·8906	0·61537
3·84	0·1432	9·1963	0·4734	0·3026	−0·8942	0·61624
3·85	0·1426	9·2795	0·4731	0·3014	−0·8978	0·61711
3·86	0·1419	9·3634	0·4728	0·3002	−0·9015	0·61797
3·87	0·1413	9·4480	0·4725	0·2990	−0·9051	0·61883
3·88	0·1407	9·5333	0·4722	0·2979	−0·9087	0·61968
3·89	0·1400	9·6192	0·4719	0·2967	−0·9123	0·62053
3·90	0·1394	9·7059	0·4716	0·2956	−0·9159	0·62137
3·91	0·1388	9·7932	0·4713	0·2944	−0·9195	0·62221
3·92	0·1382	9·8813	0·4710	0·2933	−0·9231	0·62304
3·93	0·1375	9·9701	0·4707	0·2922	−0·9267	0·62387
3·94	0·1369	10·060	0·4704	0·2911	−0·9303	0·62469
3·95	0·1363	10·150	0·4702	0·2899	−0·9339	0·62551
3·96	0·1357	10·241	0·4699	0·2888	−0·9376	0·62632
3·97	0·1351	10·332	0·4696	0·2877	−0·9411	0·62713
3·98	0·1345	10·425	0·4693	0·2866	−0·9447	0·62793
3·99	0·1339	10·518	0·4690	0·2855	−0·9483	0·62873
4·00	0·1333	10·612	0·4687	0·2844	−0·9518	0·62953

TABLE IV

FLOW OF DRY AIR THROUGH A PLANE NORMAL SHOCK WAVE

M_1	p_2/p_1	ρ_2/ρ_1	T_2/T_1	M_2	u_2/u_1	$\Delta S/c_V$	p_{02}/p_1
1·01	1·0235	1·0167	1·0067	0·9901	0·9836	0·0000	1·9171
1·02	1·0472	1·0334	1·0133	0·9805	0·9677	0·0000	1·9398
1·03	1·0711	1·0502	1·0199	0·9712	0·9522	0·0000	1·9629
1·04	1·0953	1·0670	1·0265	0·9620	0·9372	0·0000	1·9864
1·05	1·1197	1·0839	1·0330	0·9531	0·9226	0·0001	2·0103
1·06	1·1443	1·1008	1·0396	0·9445	0·9084	0·0001	2·0345
1·07	1·1692	1·1177	1·0460	0·9360	0·8947	0·0002	2·0591
1·08	1·1943	1·1347	1·0525	0·9277	0·8813	0·0002	2·0841
1·09	1·2196	1·1518	1·0589	0·9197	0·8682	0·0003	2·1094
1·10	1·2452	1·1688	1·0653	0·9118	0·8556	0·0004	2·1351
1·11	1·2710	1·1859	1·0717	0·9041	0·8432	0·0006	2·1611
1·12	1·2971	1·2031	1·0781	0·8966	0·8312	0·0007	2·1875
1·13	1·3233	1·2202	1·0845	0·8892	0·8195	0·0009	2·2142
1·14	1·3498	1·2374	1·0908	0·8821	0·8081	0·0011	2·2412
1·15	1·3766	1·2546	1·0972	0·8751	0·7970	0·0013	2·2686
1·16	1·4036	1·2719	1·1035	0·8682	0·7862	0·0016	2·2963
1·17	1·4308	1·2891	1·1099	0·8615	0·7757	0·0019	2·3243
1·18	1·4582	1·3064	1·1162	0·8549	0·7654	0·0022	2·3526
1·19	1·4859	1·3237	1·1225	0·8485	0·7554	0·0025	2·3813
1·20	1·5138	1·3410	1·1288	0·8422	0·7457	0·0029	2·4102
1·21	1·5419	1·3584	1·1351	0·8361	0·7362	0·0033	2·4395
1·22	1·5703	1·3757	1·1414	0·8300	0·7269	0·0037	2·4691
1·23	1·5989	1·3931	1·1478	0·8241	0·7178	0·0042	2·4990
1·24	1·6278	1·4104	1·1541	0·8184	0·7090	0·0047	2·5292
1·25	1·6568	1·4278	1·1604	0·8127	0·7004	0·0052	2·5597
1·26	1·6861	1·4452	1·1667	0·8072	0·6920	0·0058	2·5906
1·27	1·7157	1·4626	1·1731	0·8017	0·6837	0·0064	2·6217
1·28	1·7455	1·4800	1·1794	0·7964	0·6757	0·0070	2·6531
1·29	1·7755	1·4973	1·1858	0·7912	0·6679	0·0077	2·6848
1·30	1·8057	1·5147	1·1921	0·7861	0·6602	0·0084	2·7168
1·31	1·8362	1·5321	1·1985	0·7810	0·6527	0·0091	2·7491
1·32	1·8669	1·5495	1·2049	0·7761	0·6454	0·0099	2·7818
1·33	1·8978	1·5669	1·2112	0·7713	0·6382	0·0107	2·8147
1·34	1·9290	1·5842	1·2177	0·7665	0·6312	0·0115	2·8478
1·35	1·9604	1·6016	1·2241	0·7619	0·6244	0·0124	2·8813
1·36	1·9921	1·6189	1·2305	0·7573	0·6177	0·0133	2·9151
1·37	2·0240	1·6363	1·2369	0·7528	0·6111	0·0142	2·9492
1·38	2·0561	1·6536	1·2434	0·7484	0·6047	0·0152	2·9835
1·39	2·0884	1·6709	1·2499	0·7441	0·5985	0·0162	3·0181
1·40	2·1210	1·6882	1·2564	0·7399	0·5923	0·0172	3·0530
1·41	2·1538	1·7055	1·2629	0·7357	0·5863	0·0183	3·0882
1·42	2·1869	1·7227	1·2694	0·7316	0·5805	0·0193	3·1237
1·43	2·2201	1·7400	1·2759	0·7276	0·5747	0·0205	3·1595
1·44	2·2537	1·7572	1·2825	0·7236	0·5691	0·0216	3·1955
1·45	2·2874	1·7744	1·2891	0·7197	0·5636	0·0228	3·2319

TABLE IV

FLOW OF DRY AIR THROUGH A PLANE NORMAL SHOCK WAVE

M_1	p_2/p_1	ρ_2/ρ_1	T_2/T_1	M_2	u_2/u_1	$\Delta S/c_v$	p_{02}/p_1
1·46	2·3214	1·7916	1·2957	0·7159	0·5582	0·0241	3·2685
1·47	2·3556	1·8088	1·3023	0·7122	0·5529	0·0253	3·3054
1·48	2·3900	1·8259	1·3090	0·7085	0·5477	0·0266	3·3425
1·49	2·4247	1·8430	1·3156	0·7048	0·5426	0·0279	3·3800
1·50	2·4596	1·8601	1·3223	0·7013	0·5376	0·0293	3·4177
1·51	2·4948	1·8771	1·3290	0·6978	0·5327	0·0307	3·4557
1·52	2·5302	1·8941	1·3358	0·6943	0·5279	0·0321	3·4940
1·53	2·5658	1·9111	1·3425	0·6909	0·5233	0·0335	3·5325
1·54	2·6016	1·9281	1·3493	0·6876	0·5186	0·0350	3·5713
1·55	2·6377	1·9450	1·3561	0·6843	0·5141	0·0365	3·6104
1·56	2·6740	1·9619	1·3630	0·6811	0·5097	0·0381	3·6498
1·57	2·7106	1·9788	1·3698	0·6779	0·5054	0·0396	3·6895
1·58	2·7474	1·9956	1·3767	0·6748	0·5011	0·0412	3·7294
1·59	2·7844	2·0124	1·3836	0·6717	0·4969	0·0429	3·7696
1·60	2·8216	2·0291	1·3906	0·6687	0·4928	0·0445	3·8100
1·61	2·8591	2·0458	1·3975	0·6657	0·4888	0·0462	3·8508
1·62	2·8968	2·0625	1·4045	0·6628	0·4848	0·0479	3·8918
1·63	2·9348	2·0792	1·4115	0·6599	0·4810	0·0497	3·9331
1·64	2·9730	2·0958	1·4186	0·6570	0·4772	0·0514	3·9746
1·65	3·0114	2·1123	1·4256	0·6542	0·4734	0·0533	4·0164
1·66	3·0500	2·1288	1·4327	0·6515	0·4697	0·0551	4·0585
1·67	3·0889	2·1453	1·4399	0·6487	0·4661	0·0569	4·1009
1·68	3·1280	2·1617	1·4470	0·6461	0·4626	0·0588	4·1435
1·69	3·1674	2·1781	1·4542	0·6434	0·4591	0·0607	4·1864
1·70	3·2070	2·1944	1·4614	0·6408	0·4557	0·0627	4·2296
1·71	3·2468	2·2107	1·4686	0·6383	0·4523	0·0646	4·2730
1·72	3·2868	2·2270	1·4759	0·6357	0·4490	0·0666	4·3167
1·73	3·3271	2·2432	1·4832	0·6333	0·4458	0·0686	4·3607
1·74	3·3676	2·2593	1·4905	0·6308	0·4426	0·0707	4·4049
1·75	3·4084	2·2754	1·4979	0·6284	0·4395	0·0727	4·4494
1·76	3·4494	2·2915	1·5053	0·6260	0·4364	0·0748	4·4942
1·77	3·4906	2·3075	1·5127	0·6237	0·4334	0·0769	4·5393
1·78	3·5321	2·3235	1·5202	0·6214	0·4304	0·0791	4·5846
1·79	3·5737	2·3394	1·5277	0·6191	0·4275	0·0812	4·6301
1·80	3·6157	2·3552	1·5352	0·6168	0·4246	0·0834	4·6760
1·81	3·6578	2·3710	1·5427	0·6146	0·4218	0·0856	4·7221
1·82	3·7002	2·3868	1·5503	0·6124	0·4190	0·0878	4·7685
1·83	3·7428	2·4025	1·5579	0·6103	0·4162	0·0901	4·8151
1·84	3·7857	2·4181	1·5655	0·6081	0·4135	0·0924	4·8620
1·85	3·8288	2·4337	1·5732	0·6060	0·4109	0·0947	4·9092
1·86	3·8721	2·4493	1·5809	0·6040	0·4083	0·0970	4·9566
1·87	3·9156	2·4648	1·5886	0·6019	0·4057	0·0993	5·0043
1·88	3·9594	2·4802	1·5964	0·5999	0·4032	0·1017	5·0523
1·89	4·0035	2·4956	1·6042	0·5979	0·4007	0·1040	5·1005
1·90	4·0477	2·5109	1·6120	0·5960	0·3983	0·1065	5·1490

37

TABLE IV

FLOW OF DRY AIR THROUGH A PLANE NORMAL SHOCK WAVE

M_1	p_2/p_1	ρ_2/ρ_1	T_2/T_1	M_2	u_2/u_1	$\Delta S/c_v$	p_{02}/p_1
1·91	4·0922	2·5262	1·6199	0·5940	0·3959	0·1089	5·1978
1·92	4·1369	2·5414	1·6278	0·5921	0·3935	0·1113	5·2468
1·93	4·1819	2·5566	1·6357	0·5903	0·3911	0·1138	5·2961
1·94	4·2271	2·5717	1·6437	0·5884	0·3888	0·1163	5·3456
1·95	4·2725	2·5867	1·6517	0·5866	0·3866	0·1188	5·3955
1·96	4·3182	2·6017	1·6597	0·5848	0·3844	0·1213	5·4455
1·97	4·3640	2·6167	1·6678	0·5830	0·3822	0·1239	5·4959
1·98	4·4102	2·6315	1·6759	0·5812	0·3800	0·1264	5·5465
1·99	4·4565	2·6464	1·6840	0·5795	0·3779	0·1290	5·5974
2·00	4·5031	2·6611	1·6922	0·5778	0·3758	0·1316	5·6485
2·01	4·5499	2·6758	1·7004	0·5761	0·3737	0·1342	5·6999
2·02	4·5970	2·6905	1·7086	0·5744	0·3717	0·1368	5·7516
2·03	4·6443	2·7051	1·7169	0·5727	0·3697	0·1395	5·8035
2·04	4·6918	2·7196	1·7252	0·5711	0·3677	0·1421	5·8557
2·05	4·7396	2·7341	1·7335	0·5695	0·3658	0·1448	5·9081
2·06	4·7876	2·7485	1·7419	0·5679	0·3638	0·1475	5·9608
2·07	4·8358	2·7628	1·7503	0·5663	0·3619	0·1502	6·0138
2·08	4·8843	2·7771	1·7587	0·5648	0·3601	0·1530	6·0670
2·09	4·9330	2·7914	1·7672	0·5632	0·3582	0·1557	6·1205
2·10	4·9819	2·8056	1·7757	0·5617	0·3564	0·1585	6·1743
2·11	5·0310	2·8197	1·7843	0·5602	0·3547	0·1612	6·2283
2·12	5·0804	2·8337	1·7928	0·5587	0·3529	0·1640	6·2826
2·13	5·1301	2·8477	1·8015	0·5573	0·3512	0·1668	6·3371
2·14	5·1799	2·8617	1·8101	0·5558	0·3494	0·1697	6·3919
2·15	5·2300	2·8755	1·8188	0·5544	0·3478	0·1725	6·4470
2·16	5·2803	2·8894	1·8275	0·5530	0·3461	0·1754	6·5023
2·17	5·3309	2·9031	1·8363	0·5516	0·3445	0·1782	6·5579
2·18	5·3817	2·9168	1·8451	0·5502	0·3428	0·1811	6·6138
2·19	5·4327	2·9305	1·8539	0·5489	0·3412	0·1840	6·6699
2·20	5·4840	2·9440	1·8627	0·5475	0·3397	0·1869	6·7263
2·21	5·5355	2·9576	1·8716	0·5462	0·3381	0·1898	6·7829
2·22	5·5872	2·9710	1·8806	0·5449	0·3366	0·1927	6·8398
2·23	5·6392	2·9844	1·8895	0·5436	0·3351	0·1957	6·8970
2·24	5·6914	2·9978	1·8985	0·5423	0·3336	0·1986	6·9544
2·25	5·7438	3·0110	1·9076	0·5410	0·3321	0·2016	7·0121
2·26	5·7965	3·0243	1·9167	0·5398	0·3307	0·2046	7·0700
2·27	5·8494	3·0374	1·9258	0·5385	0·3292	0·2076	7·1282
2·28	5·9025	3·0505	1·9349	0·5373	0·3278	0·2106	7·1867
2·29	5·9559	3·0636	1·9441	0·5361	0·3264	0·2136	7·2454
2·30	6·0095	3·0765	1·9533	0·5349	0·3250	0·2166	7·3044
2·31	6·0633	3·0895	1·9626	0·5337	0·3237	0·2197	7·3637
2·32	6·1174	3·1023	1·9719	0·5326	0·3223	0·2227	7·4232
2·33	6·1717	3·1151	1·9812	0·5314	0·3210	0·2258	7·4829
2·34	6·2262	3·1279	1·9906	0·5303	0·3197	0·2289	7·5430
2·35	6·2810	3·1405	2·0000	0·5291	0·3184	0·2319	7·6033

TABLE IV

FLOW OF DRY AIR THROUGH A PLANE NORMAL SHOCK WAVE

M_1	p_2/p_1	ρ_2/ρ_1	T_2/T_1	M_2	u_2/u_1	$\Delta S/c_V$	p_{02}/p_1
2·36	6·3360	3·1532	2·0094	0·5280	0·3171	0·2350	7·6638
2·37	6·3912	3·1657	2·0189	0·5269	0·3159	0·2381	7·7246
2·38	6·4467	3·1782	2·0284	0·5258	0·3146	0·2412	7·7857
2·39	6·5024	3·1907	2·0379	0·5247	0·3134	0·2444	7·8470
2·40	6·5583	3·2031	2·0475	0·5236	0·3122	0·2475	7·9086
2·41	6·6145	3·2154	2·0571	0·5226	0·3110	0·2506	7·9705
2·42	6·6709	3·2276	2·0668	0·5215	0·3098	0·2538	8·0326
2·43	6·7275	3·2398	2·0765	0·5205	0·3087	0·2569	8·0950
2·44	6·7844	3·2520	2·0862	0·5195	0·3075	0·2601	8·1576
2·45	6·8415	3·2641	2·0960	0·5185	0·3064	0·2633	8·2205
2·46	6·8988	3·2761	2·1058	0·5175	0·3052	0·2665	8·2837
2·47	6·9564	3·2881	2·1156	0·5165	0·3041	0·2697	8·3471
2·48	7·0142	3·3000	2·1255	0·5155	0·3030	0·2729	8·4107
2·49	7·0722	3·3119	2·1354	0·5145	0·3019	0·2761	8·4747
2·50	7·1305	3·3237	2·1454	0·5135	0·3009	0·2793	8·5389
2·51	7·1890	3·3354	2·1554	0·5126	0·2998	0·2825	8·6033
2·52	7·2477	3·3471	2·1654	0·5116	0·2988	0·2857	8·6680
2·53	7·3067	3·3587	2·1754	0·5107	0·2977	0·2890	8·7330
2·54	7·3659	3·3703	2·1855	0·5098	0·2967	0·2922	8·7983
2·55	7·4253	3·3818	2·1957	0·5089	0·2957	0·2955	8·8637
2·56	7·4850	3·3932	2·2059	0·5080	0·2947	0·2987	8·9295
2·57	7·5449	3·4046	2·2161	0·5071	0·2937	0·3020	8·9955
2·58	7·6050	3·4160	2·2263	0·5062	0·2927	0·3053	9·0618
2·59	7·6654	3·4272	2·2366	0·5053	0·2918	0·3086	9·1283
2·60	7·7260	3·4385	2·2469	0·5044	0·2908	0·3118	9·1951
2·61	7·7868	3·4496	2·2573	0·5036	0·2899	0·3151	9·2622
2·62	7·8479	3·4607	2·2677	0·5027	0·2890	0·3184	9·3295
2·63	7·9092	3·4718	2·2781	0·5019	0·2880	0·3217	9·3970
2·64	7·9707	3·4828	2·2886	0·5011	0·2871	0·3251	9·4649
2·65	8·0325	3·4938	2·2991	0·5002	0·2862	0·3284	9·5330
2·66	8·0945	3·5046	2·3097	0·4994	0·2853	0·3317	9·6013
2·67	8·1568	3·5155	2·3202	0·4986	0·2845	0·3350	9·6699
2·68	8·2192	3·5263	2·3309	0·4978	0·2836	0·3384	9·7388
2·69	8·2819	3·5370	2·3415	0·4970	0·2827	0·3417	9·8079
2·70	8·3449	3·5477	2·3522	0·4962	0·2819	0·3450	9·8773
2·71	8·4081	3·5583	2·3630	0·4955	0·2810	0·3484	9·9469
2·72	8·4715	3·5688	2·3737	0·4947	0·2802	0·3518	10·0169
2·73	8·5351	3·5793	2·3845	0·4939	0·2794	0·3551	10·0870
2·74	8·5990	3·5898	2·3954	0·4932	0·2786	0·3585	10·1574
2·75	8·6631	3·6002	2·4063	0·4924	0·2778	0·3618	10·2281
2·76	8·7274	3·6105	2·4172	0·4917	0·2770	0·3652	10·2991
2·77	8·7920	3·6208	2·4282	0·4909	0·2762	0·3686	10·3703
2·78	8·8568	3·6311	2·4392	0·4902	0·2754	0·3720	10·4417
2·79	8·9218	3·6413	2·4502	0·4895	0·2746	0·3754	10·5135
2·80	8·9871	3·6514	2·4613	0·4888	0·2739	0·3787	10·5854

TABLE IV

FLOW OF DRY AIR THROUGH A PLANE NORMAL SHOCK WAVE

M_1	p_2/p_1	ρ_2/ρ_1	T_2/T_1	M_2	u_2/u_1	$\Delta S/c_v$	p_{02}/p_1
2·81	9·0526	3·6615	2·4724	0·4881	0·2731	0·3821	10·6577
2·82	9·1184	3·6715	2·4835	0·4874	0·2724	0·3855	10·7302
2·83	9·1843	3·6815	2·4947	0·4867	0·2716	0·3889	10·8029
2·84	9·2506	3·6914	2·5060	0·4860	0·2709	0·3923	10·8759
2·85	9·3170	3·7013	2·5172	0·4853	0·2702	0·3958	10·9492
2·86	9·3837	3·7111	2·5285	0·4846	0·2695	0·3992	11·0227
2·87	9·4506	3·7209	2·5399	0·4840	0·2688	0·4026	11·0965
2·88	9·5177	3·7306	2·5512	0·4833	0·2681	0·4060	11·1706
2·89	9·5851	3·7403	2·5627	0·4827	0·2674	0·4094	11·2449
2·90	9·6527	3·7499	2·5741	0·4820	0·2667	0·4128	11·3195
2·91	9·7206	3·7595	2·5856	0·4814	0·2660	0·4163	11·3943
2·92	9·7886	3·7690	2·5971	0·4807	0·2653	0·4197	11·4694
2·93	9·8569	3·7785	2·6087	0·4801	0·2647	0·4231	11·5447
2·94	9·9255	3·7879	2·6203	0·4795	0·2640	0·4266	11·6203
2·95	9·9943	3·7973	2·6319	0·4789	0·2633	0·4300	11·6962
2·96	10·0633	3·8066	2·6436	0·4782	0·2627	0·4334	11·7723
2·97	10·1325	3·8159	2·6553	0·4776	0·2621	0·4369	11·8487
2·98	10·2020	3·8251	2·6671	0·4770	0·2614	0·4403	11·9254
2·99	10·2717	3·8343	2·6789	0·4764	0·2608	0·4438	12·0023
3·00	10·3417	3·8434	2·6907	0·4758	0·2602	0·4472	12·0794
3·01	10·4118	3·8525	2·7026	0·4753	0·2596	0·4507	12·1568
3·02	10·4822	3·8616	2·7145	0·4747	0·2590	0·4541	12·2345
3·03	10·5529	3·8705	2·7265	0·4741	0·2584	0·4576	12·3125
3·04	10·6238	3·8795	2·7385	0·4735	0·2578	0·4610	12·3906
3·05	10·6949	3·8884	2·7505	0·4730	0·2572	0·4645	12·4091
3·06	10·7662	3·8972	2·7625	0·4724	0·2566	0·4680	12·5478
3·07	10·8378	3·9060	2·7746	0·4718	0·2560	0·4714	12·6268
3·08	10·9096	3·9148	2·7868	0·4713	0·2554	0·4749	12·7060
3·09	10·9817	3·9235	2·7990	0·4707	0·2549	0·4784	12·7855
3·10	11·0540	3·9321	2·8112	0·4702	0·2543	0·4818	12·8653
3·11	11·1265	3·9408	2·8234	0·4697	0·2538	0·4853	12·9453
3·12	11·1992	3·9493	2·8357	0·4691	0·2532	0·4888	13·0255
3·13	11·2722	3·9579	2·8481	0·4686	0·2527	0·4922	13·1060
3·14	11·3454	3·9663	2·8604	0·4681	0·2521	0·4957	13·1868
3·15	11·4189	3·9748	2·8728	0·4676	0·2516	0·4992	13·2679
3·16	11·4925	3·9832	2·8853	0·4671	0·2511	0·5026	13·3492
3·17	11·5665	3·9915	2·8978	0·4665	0·2505	0·5061	13·4307
3·18	11·6406	3·9998	2·9103	0·4660	0·2500	0·5096	13·5125
3·19	11·7150	4·0081	2·9228	0·4655	0·2495	0·5131	13·5946
3·20	11·7896	4·0163	2·9354	0·4650	0·2490	0·5165	13·6770
3·21	11·8645	4·0245	2·9481	0·4645	0·2485	0·5200	13·7595
3·22	11·9395	4·0326	2·9608	0·4641	0·2480	0·5235	13·8424
3·23	12·0149	4·0407	2·9735	0·4636	0·2475	0·5270	13·9255
3·24	12·0904	4·0487	2·9862	0·4631	0·2470	0·5305	14·0089
3·25	12·1662	4·0567	2·9990	0·4626	0·2465	0·5339	14·0925

TABLE IV

FLOW OF DRY AIR THROUGH A PLANE NORMAL SHOCK WAVE

M_1	p_2/p_1	ρ_2/ρ_1	T_2/T_1	M_2	u_2/u_1	$\Delta S/c_v$	p_{02}/p_1
3·26	12·2422	4·0647	3·0118	0·4621	0·2460	0·5374	14·1764
3·27	12·3185	4·0726	3·0247	0·4617	0·2455	0·5409	14·2605
3·28	12·3950	4·0805	3·0376	0·4612	0·2451	0·5444	14·3449
3·29	12·4717	4·0883	3·0506	0·4607	0·2446	0·5478	14·4296
3·30	12·5486	4·0961	3·0635	0·4603	0·2441	0·5513	14·5145
3·31	12·6258	4·1039	3·0766	0·4598	0·2437	0·5548	14·5997
3·32	12·7032	4·1116	3·0896	0·4594	0·2432	0·5583	14·6851
3·33	12·7809	4·1192	3·1027	0·4589	0·2428	0·5618	14·7708
3·34	12·8588	4·1269	3·1159	0·4585	0·2423	0·5652	14·8568
3·35	12·9369	4·1345	3·1290	0·4581	0·2419	0·5687	14·9430
3·36	13·0152	4·1420	3·1423	0·4576	0·2414	0·5722	15·0295
3·37	13·0938	4·1495	3·1555	0·4572	0·2410	0·5757	15·1162
3·38	13·1726	4·1570	3·1688	0·4568	0·2406	0·5792	15·2032
3·39	13·2517	4·1644	3·1821	0·4563	0·2401	0·5826	15·2904
3·40	13·3310	4·1718	3·1955	0·4559	0·2397	0·5861	15·3779
3·41	13·4105	4·1792	3·2089	0·4555	0·2393	0·5896	15·4657
3·42	13·4903	4·1865	3·2224	0·4551	0·2389	0·5931	15·5537
3·43	13·5702	4·1937	3·2358	0·4547	0·2385	0·5965	15·6420
3·44	13·6505	4·2010	3·2494	0·4543	0·2380	0·6000	15·7306
3·45	13·7309	4·2082	3·2629	0·4539	0·2376	0·6035	15·8193
3·46	13·8116	4·2153	3·2765	0·4535	0·2372	0·6070	15·9084
3·47	13·8925	4·2225	3·2902	0·4531	0·2368	0·6104	15·9977
3·48	13·9737	4·2295	3·3038	0·4527	0·2364	0·6139	16·0873
3·49	14·0551	4·2366	3·3176	0·4523	0·2360	0·6174	16·1771
3·50	14·1367	4·2436	3·3313	0·4519	0·2356	0·6209	16·2672
3·51	14·2186	4·2506	3·3451	0·4515	0·2353	0·6243	16·3576
3·52	14·3007	4·2575	3·3589	0·4511	0·2349	0·6278	16·4482
3·53	14·3830	4·2644	3·3728	0·4507	0·2345	0·6313	16·5390
3·54	14·4655	4·2713	3·3867	0·4504	0·2341	0·6347	16·6302
3·55	14·5483	4·2781	3·4007	0·4500	0·2337	0·6382	16·7215
3·56	14·6313	4·2849	3·4146	0·4496	0·2334	0·6417	16·8132
3·57	14·7146	4·2916	3·4287	0·4492	0·2330	0·6451	16·9051
3·58	14·7981	4·2984	3·4427	0·4489	0·2326	0·6486	16·9972
3·59	14·8818	4·3050	3·4568	0·4485	0·2323	0·6521	17·0897
3·60	14·9658	4·3117	3·4710	0·4482	0·2319	0·6555	17·1823
3·61	15·0500	4·3183	3·4851	0·4478	0·2316	0·6590	17·2753
3·62	15·1344	4·3249	3·4994	0·4474	0·2312	0·6624	17·3684
3·63	15·2191	4·3314	3·5136	0·4471	0·2309	0·6659	17·4619
3·64	15·3039	4·3380	3·5279	0·4467	0·2305	0·6693	17·5556
3·65	15·3891	4·3444	3·5423	0·4464	0·2302	0·6728	17·6496
3·66	15·4744	4·3509	3·5566	0·4461	0·2298	0·6763	17·7438
3·67	15·5600	4·3573	3·5710	0·4457	0·2295	0·6797	17·8383
3·68	15·6458	4·3637	3·5855	0·4454	0·2292	0·6831	17·9330
3·69	15·7319	4·3700	3·6000	0·4450	0·2288	0·6866	18·0280
3·70	15·8182	4·3763	3·6145	0·4447	0·2285	0·6900	18·1233

TABLE IV

FLOW OF DRY AIR THROUGH A PLANE NORMAL SHOCK WAVE

M_1	p_2/p_1	ρ_2/ρ_1	T_2/T_1	M_2	u_2/u_1	$\Delta S/c_v$	p_{02}/p_1
3·71	15·9047	4·3826	3·6291	0·4444	0·2282	0·6935	18·2188
3·72	15·9915	4·3888	3·6437	0·4440	0·2279	0·6969	18·3146
3·73	16·0785	4·3950	3·6583	0·4437	0·2275	0·7004	18·4106
3·74	16·1657	4·4012	3·6730	0·4434	0·2272	0·7038	18·5069
3·75	16·2532	4·4074	3·6877	0·4431	0·2269	0·7072	18·6034
3·76	16·3409	4·4135	3·7025	0·4428	0·2266	0·7107	18·7002
3·77	16·4288	4·4196	3·7173	0·4424	0·2263	0·7141	18·7973
3·78	16·5170	4·4256	3·7321	0·4421	0·2260	0·7175	18·8946
3·79	16·6054	4·4316	3·7470	0·4418	0·2256	0·7210	18·9922
3·80	16·6940	4·4376	3·7619	0·4415	0·2253	0·7244	19·0901
3·81	16·7828	4·4436	3·7769	0·4412	0·2250	0·7278	19·1882
3·82	16·8719	4·4495	3·7919	0·4409	0·2247	0·7313	19·2865
3·83	16·9613	4·4554	3·8069	0·4406	0·2244	0·7347	19·3851
3·84	17·0508	4·4613	3·8220	0·4403	0·2242	0·7381	19·4840
3·85	17·1406	4·4671	3·8371	0·4400	0·2239	0·7415	19·5831
3·86	17·2307	4·4729	3·8522	0·4397	0·2236	0·7449	19·6825
3·87	17·3209	4·4787	3·8674	0·4394	0·2233	0·7483	19·7822
3·88	17·4114	4·4845	3·8826	0·4391	0·2230	0·7518	19·8821
3·89	17·5022	4·4902	3·8979	0·4388	0·2227	0·7552	19·9822
3·90	17·5931	4·4959	3·9132	0·4385	0·2224	0·7586	20·0827
3·91	17·6843	4·5015	3·9285	0·4382	0·2221	0·7620	20·1833
3·92	17·7757	4·5071	3·9439	0·4379	0·2219	0·7654	20·2843
3·93	17·8674	4·5127	3·9593	0·4377	0·2216	0·7688	20·3855
3·94	17·9593	4·5183	3·9748	0·4374	0·2213	0·7722	20·4869
3·95	18·0514	4·5239	3·9903	0·4371	0·2211	0·7756	20·5886
3·96	18·1438	4·5294	4·0058	0·4368	0·2208	0·7790	20·6906
3·97	18·2364	4·5348	4·0214	0·4366	0·2205	0·7824	20·7928
3·98	18·3292	4·5403	4·0370	0·4363	0·2202	0·7858	20·8953
3·99	18·4223	4·5457	4·0527	0·4360	0·2200	0·7892	20·9981
4·00	18·5156	4·5511	4·0683	0·4357	0·2197	0·7925	21·1022

TABLE V

FLOW OF DRY AIR THROUGH A PLANE OBLIQUE SHOCK WAVE

M_1	δ	β	p_2/p_1	ρ_2/ρ_1	T_2/T_1	M_2	$\Delta S/c_v$
1·05	0·557	79·937	1·0804	1·0566	1·0225	0·9845	0·00000
1·10	1·513	76·296	1·1659	1·1155	1·0452	0·9711	0·00017
1·15	2·000	67·017	1·1412	1·0986	1·0387	1·0432	0·00012
	2·667	73·821	1·2567	1·1765	1·0682	0·9598	0·00047
	2·000	81·155	1·3401	1·2311	1·0885	0·9008	0·00105
1·20	2·000	61·058	1·1200	1·0841	1·0331	1·1112	0·00006
	3·938	71·975	1·3528	1·2393	1·0915	0·9503	0·00111
	2·000	83·850	1·4945	1·3291	1·1245	0·8552	0·00262
1·25	2·000	56·850	1·1113	1·0781	1·0308	1·1694	0·00006
	4·000	62·006	1·2548	1·1752	1·0677	1·0717	0·00047
	5·278	70·537	1·4543	1·3040	1·1153	0·9424	0·00216
	4·000	79·360	1·5946	1·3905	1·1468	0·8527	0·00414
	2·000	85·202	1·6441	1·4202	1·1576	0·8209	0·00501
1·30	2·000	53·478	1·1068	1·0750	1·0296	1·2242	0·00006
	4·000	57·435	1·2340	1·1614	1·0625	1·1395	0·00041
	6·000	63·499	1·4128	1·2778	1·1057	1·0267	0·00169
	6·651	69·392	1·5612	1·3702	1·1394	0·9359	0·00362
	6·000	75·324	1·6791	1·4410	1·1652	0·8641	0·00566
	4·000	81·631	1·7639	1·4907	1·1833	0·8120	0·00746
	2·000	86·050	1·7964	1·5094	1·1901	0·7919	0·00816
1·35	2·000	50·637	1·1044	1·0733	1·0290	1·2773	0·00006
	4·000	53·975	1·2243	1·1549	1·0601	1·1991	0·00035
	6·000	58·254	1·3713	1·2512	1·0959	1·1084	0·00128
	8·000	67·143	1·6393	1·4174	1·1566	0·9507	0·00490
	8·035	68·466	1·6737	1·4378	1·1641	0·9307	0·00554
	8·000	69·786	1·7064	1·4571	1·1711	0·9118	0·00624
	6·000	78·631	1·8777	1·5556	1·2071	0·8114	0·01015
	4·000	83·013	1·9290	1·5842	1·2176	0·7809	0·01149
	2·000	86·637	1·9531	1·5975	1·2226	0·7663	0·01209
1·40	2·000	48·177	1·1033	1·0725	1·0287	1·3293	0·00006
	4·000	51·126	1·2195	1·1517	1·0589	1·2550	0·00029
	6·000	54·650	1·3549	1·2407	1·0920	1·1732	0·00117
	8·000	59·404	1·5281	1·3499	1·1320	1·0736	0·00309
	9·411	67·710	1·7917	1·5067	1·1892	0·9267	0·00805
	8·000	75·849	1·9842	1·6146	1·2289	0·8189	0·01306
	6·000	80·461	2·0581	1·6547	1·2438	0·7765	0·01522
	4·000	83·974	2·0958	1·6748	1·2513	0·7547	0·01638
	2·000	87·069	2·1150	1·6850	1·2552	0·7434	0·01697
1·45	2·000	46·007	1·1030	1·0723	1·0286	1·3806	0·00006
	4·000	48·687	1·2174	1·1503	1·0584	1·3088	0·00029
	6·000	51·770	1·3472	1·2357	1·0902	1·2320	0·00111
	8·000	55·542	1·5015	1·3334	1·1260	1·1453	0·00274
	10·000	61·110	1·7143	1·4618	1·1728	1·0303	0·00636
	10·766	67·091	1·9154	1·5766	1·2148	0·9236	0·01114
	10·000	72·922	2·0757	1·6641	1·2473	0·8376	0·01574

TABLE V

FLOW OF DRY AIR THROUGH A PLANE OBLIQUE SHOCK WAVE

M_1	δ	β	p_2/p_1	ρ_2/ρ_1	T_2/T_1	M_2	$\Delta S/c_v$
1·45	8·000	78·164	2·1841	1·7213	1·2689	0·7781	0·01924
	6·000	81·713	2·2364	1·7484	1·2791	0·7488	0·02105
	4·000	84·689	2·2664	1·7637	1·2850	0·7318	0·02210
	2·000	87·400	2·2823	1·7718	1·2881	0·7226	0·02268
1·50	2·000	44·068	1·1032	1·0725	1·0286	1·4315	0·00005
	4·000	46·550	1·2170	1·1500	1·0583	1·3612	0·00029
	6·000	49·339	1·3442	1·2337	1·0895	1·2875	0·00105
	8·000	52·592	1·4901	1·3263	1·1234	1·2072	0·00257
	10·000	56·716	1·6684	1·4347	1·1629	1·1134	0·00548
	12·000	64·580	1·9755	1·6099	1·2271	0·9565	0·01277
	12·091	66·581	2·0446	1·6474	1·2411	0·9213	0·01481
	12·000	68·559	2·1086	1·6816	1·2539	0·8886	0·01679
	10·000	75·949	2·3048	1·7832	1·2925	0·7860	0·02344
	8·000	79·683	2·3754	1·8186	1·3061	0·7479	0·02606
	6·000	82·643	2·4166	1·8390	1·3141	0·7253	0·02764
	4·000	85·244	2·4416	1·8513	1·3189	0·7114	0·02857
	2·000	87·662	2·4553	1·8579	1·3215	0·7037	0·02910
1·55	2·000	42·318	1·1039	1·0729	1·0288	1·4819	0·00006
	4·000	44·648	1·2178	1·1505	1·0585	1·4127	0·00029
	6·000	47·226	1·3439	1·2336	1·0894	1·3409	0·00105
	8·000	50·149	1·4858	1·3236	1·1225	1·2645	0·00251
	10·000	53·628	1·6511	1·4244	1·1592	1·1795	0·00513
	12·000	58·298	1·8630	1·5473	1·2040	1·0744	0·00980
	13·378	66·162	2·1795	1·7189	1·2679	0·9198	0·01913
	12·000	73·621	2·4146	1·8380	1·3137	0·8023	0·02752
	10·000	77·766	2·5117	1·8853	1·3323	0·7520	0·03137
	8·000	80·800	2·5660	1·9112	1·3426	0·7233	0·03353
	6·000	83·368	2·6003	1·9275	1·3491	0·7048	0·03499
	4·000	85·688	2·6218	1·9376	1·3531	0·6930	0·03586
	2·000	87·874	2·6338	1·9432	1·3554	0·6864	0·03638
1·60	2·000	40·727	1·1048	1·0736	1·0291	1·5321	0·00006
	4·000	42·937	1·2194	1·1516	1·0589	1·4635	0·00029
	6·000	45·354	1·3454	1·2346	1·0898	1·3929	0·00105
	8·000	48·047	1·4856	1·3236	1·1224	1·3189	0·00251
	10·000	51·140	1·6449	1·4207	1·1578	1·2388	0·00501
	12·000	54·930	1·8347	1·5313	1·1982	1·1471	0·00910
	14·000	60·639	2·1030	1·6786	1·2528	1·0208	0·01662
	14·624	65·818	2·3200	1·7909	1·2954	0·9189	0·02402
	14·000	70·783	2·4978	1·8786	1·3296	0·8337	0·03079
	12·000	75·850	2·6430	1·9475	1·3571	0·7618	0·03673
	10·000	79·069	2·7141	1·9804	1·3705	0·7255	0·03983
	8·000	81·668	2·7589	2·0008	1·3789	0·7022	0·04175
	6·000	83·951	2·7884	2·0142	1·3844	0·6865	0·04303
	4·000	86·051	2·8074	2·0228	1·3879	0·6763	0·04391
	2·000	88·049	2·8182	2·0276	1·3899	0·6705	0·04437
1·65	2·000	39·270	1·1060	1·0744	1·0294	1·5821	0·00006
	4·000	41·383	1·2217	1·1531	1·0594	1·5137	0·0003
	6·000	43·675	1·3484	1·2365	1·0905	1·4439	0·0011
	8·000	46·196	1·4882	1·3252	1·1230	1·3714	0·0026

TABLE V

FLOW OF DRY AIR THROUGH A PLANE OBLIQUE SHOCK WAVE

M_1	δ	β	p_2/p_1	ρ_2/ρ_1	T_2/T_1	M_2	$\Delta S/c_v$
1·65	10·000	49·030	1·6447	1·4206	1·1578	1·2943	0·0050
	12·000	52·345	1·8249	1·5257	1·1961	1·2093	0·0089
	14·000	56·600	2·0480	1·6493	1·2418	1·1073	0·0149
	15·825	65·536	2·4662	1·8633	1·3236	0·9185	0·0296
	14·000	73·795	2·7638	2·0031	1·3798	0·7792	0·0420
	12·000	77·369	2·8594	2·0460	1·3976	0·7323	0·0462
	10·000	80·072	2·9169	2·0713	1·4082	0·7033	0·0489
	8·000	82·368	2·9553	2·0881	1·4153	0·6836	0·0506
	6·000	84·431	2·9814	2·0994	1·4201	0·6700	0·0518
	4·000	86·354	2·9985	2·1068	1·4233	0·6610	0·0527
	2·000	88·196	3·0082	2·1110	1·4251	0·6559	0·0531
1·70	2·000	37·930	1·1074	1·0754	1·0298	1·6319	0·0001
	4·000	39·962	1·2244	1·1550	1·0601	1·5635	0·0003
	6·000	42·154	1·3523	1·2390	1·0914	1·4940	0·0011
	8·000	44·542	1·4927	1·3280	1·1240	1·4225	0·0026
	10·000	47·187	1·6483	1·4227	1·1586	1·3473	0·0051
	12·000	50·198	1·8241	1·5252	1·1960	1·2662	0·0088
	14·000	53·817	2·0308	1·6399	1·2383	1·1742	0·0144
	16·000	58·888	2·3060	1·7838	1·2927	1·0545	0·0235
	16·979	65·307	2·6180	1·9358	1·3524	0·9185	0·0357
	16·000	71·322	2·8609	2·0466	1·3978	0·8093	0·0463
	14·000	75·615	2·9987	2·1069	1·4233	0·7447	0·0527
	12·000	78·517	3·0732	2·1387	1·4370	0·7085	0·0562
	10·000	80·878	3·1221	2·1593	1·4459	0·6842	0·0585
	8·000	82·945	3·1561	2·1734	1·4521	0·6671	0·0602
	6·000	84·833	3·1796	2·1832	1·4564	0·6550	0·0613
	4·000	86·610	3·1952	2·1896	1·4593	0·6470	0·0621
	2·000	88·321	3·2041	2·1932	1·4609	0·6423	0·0625
1·75	2·000	36·691	1·1090	1·0765	1·0302	1·6815	0·0001
	4·000	38·656	1·2276	1·1571	1·0609	1·6129	0·0003
	6·000	40·765	1·3570	1·2420	1·0925	1·5436	0·0012
	8·000	43·048	1·4986	1·3316	1·1254	1·4725	0·0027
	10·000	45·549	1·6546	1·4265	1·1599	1·3986	0·0052
	12·000	48·346	1·8287	1·5278	1·1969	1·3199	0·0089
	14·000	51·586	2·0278	1·6383	1·2377	1·2334	0·0143
	16·000	55·653	2·2700	1·7656	1·2857	1·1310	0·0222
	18·000	63·287	2·6858	1·9674	1·3652	0·9570	0·0386
	18·085	65·121	2·7755	2·0084	1·3820	0·9189	0·0425
	18·000	66·917	2·8587	2·0457	1·3974	0·8832	0·0462
	16·000	73·683	3·1261	2·1609	1·4467	0·7646	0·0587
	14·000	76·940	3·2258	2·2022	1·4648	0·7182	0·0636
	12·000	79·430	3·2881	2·2275	1·4761	0·6884	0·0667
	10·000	81·543	3·3310	2·2447	1·4839	0·6673	0·0688
	8·000	83·431	3·3616	2·2569	1·4895	0·6521	0·0704
	6·000	85·176	3·3831	2·2655	1·4933	0·6413	0·0714
	4·000	86·829	3·3975	2·2711	1·4959	0·6340	0·0722
	2·000	88·428	3·4057	2·2744	1·4974	0·6298	0·0726
1·80	2·000	35·541	1·1106	1·0776	1·0306	1·7309	0·0001
	4·000	37·448	1·2311	1·1594	1·0618	1·6621	0·0003
	6·000	39·489	1·3623	1·2455	1·0938	1·5926	0·0012

TABLE V

FLOW OF DRY AIR THROUGH A PLANE OBLIQUE SHOCK WAVE

M_1	δ	β	p_2/p_1	ρ_2/ρ_1	T_2/T_1	M_2	$\Delta S/c_V$
1·80	8·000	41·686	1·5056	1·3360	1·1270	1·5218	0·0028
	10·000	44·075	1·6629	1·4314	1·1617	1·4485	0·0054
	12·000	46·711	1·8369	1·5325	1·1986	1·3714	0·0092
	14·000	49·696	2·0327	1·6410	1·2387	1·2882	0·0145
	16·000	53·250	2·2612	1·7611	1·2840	1·1940	0·0219
	18·000	58·094	2·5588	1·9078	1·3412	1·0738	0·0332
	19·145	64·973	2·9386	2·0808	1·4122	0·9196	0·0499
	18·000	71·314	3·2273	2·2028	1·4651	0·7975	0·0637
	16·000	75·262	3·3708	2·2606	1·4911	0·7337	0·0708
	14·000	77·976	3·4515	2·2923	1·5057	0·6965	0·0749
	12·000	80·182	3·5056	2·3133	1·5154	0·6709	0·0777
	10·000	82·103	3·5442	2·3281	1·5224	0·6522	0·0797
	8·000	83·845	3·5722	2·3388	1·5274	0·6385	0·0812
	6·000	85·471	3·5921	2·3463	1·5309	0·6287	0·0822
	4·000	87·019	3·6054	2·3514	1·5333	0·6220	0·0829
	2·000	88·521	3·6131	2·3543	1·5347	0·6181	0·0833
1·85	2·000	34·469	1·1124	1·0789	1·0311	1·7803	0·0001
	4·000	36·328	1·2348	1·1619	1·0628	1·7110	0·0004
	6·000	38·310	1·3682	1·2492	1·0952	1·6412	0·0013
	8·000	40·437	1·5136	1·3409	1·1288	1·5703	0·0029
	10·000	42·734	1·6727	1·4372	1·1638	1·4974	0·0055
	12·000	45·246	1·8477	1·5386	1·2009	1·4212	0·0095
	14·000	48·047	2·0426	1·6464	1·2407	1·3401	0·0148
	16·000	51·277	2·2649	1·7630	1·2847	1·2507	0·0220
	18·000	55·299	2·5336	1·8958	1·3364	1·1453	0·0322
	20·000	62·392	2·9705	2·0947	1·4181	0·9748	0·0514
	20·157	64·856	3·1073	2·1530	1·4432	0·9205	0·0578
	20·000	67·242	3·2307	2·2042	1·4657	0·8707	0·0638
	18·000	73·358	3·5010	2·3115	1·5146	0·7573	0·0775
	16·000	76·457	3·6096	2·3529	1·5341	0·7094	0·0831
	14·000	78·820	3·6785	2·3788	1·5464	0·6780	0·0867
	12·000	80·813	3·7269	2·3966	1·5551	0·6554	0·0893
	10·000	82·582	3·7621	2·4096	1·5613	0·6386	0·0911
	8·000	84·204	3·7880	2·4190	1·5659	0·6261	0·0925
	6·000	85·727	3·8066	2·4257	1·5693	0·6170	0·0935
	4·000	87·185	3·8191	2·4303	1·5715	0·6108	0·0942
	2·000	88·602	3·8264	2·4329	1·5728	0·6072	0·0945
1·90	2·000	33·468	1·1143	1·0801	1·0316	1·8296	0·0001
	4·000	35·284	1·2388	1·1646	1·0637	1·7596	0·0004
	6·000	37·217	1·3744	1·2533	1·0967	1·6894	0·0013
	8·000	39·284	1·5222	1·3463	1·1307	1·6183	0·0030
	10·000	41·507	1·6836	1·4437	1·1662	1·5454	0·0058
	12·000	43·920	1·8606	1·5459	1·2035	1·4697	0·0097
	14·000	46·580	2·0562	1·6536	1·2434	1·3899	0·0152
	16·000	49·585	2·2759	1·7686	1·2869	1·3035	0·0224
	18·000	53·155	2·5319	1·8950	1·3361	1·2056	0·0322
	20·000	58·014	2·8649	2·0484	1·3986	1·0802	0·0465
	21·124	64·766	3·2816	2·2249	1·4750	0·9216	0·0664
	20·000	70·934	3·5979	2·3485	1·5320	0·7957	0·0825
	18·000	74·792	3·7576	2·4079	1·5605	0·7285	0·0909
	16·000	77·413	3·8475	2·4405	1·5765	0·6892	0·0957

TABLE V

FLOW OF DRY AIR THROUGH A PLANE OBLIQUE SHOCK WAVE

M_1	δ	β	p_2/p_1	ρ_2/ρ_1	T_2/T_1	M_2	$\Delta S/c_V$
1·90	14·000	79·526	3·9084	2·4622	1·5874	0·6617	0·0990
	12·000	81·353	3·9524	2·4778	1·5952	0·6414	0·1013
	10·000	82·996	3·9850	2·4892	1·6009	0·6261	0·1031
	8·000	84·516	4·0092	2·4976	1·6052	0·6146	0·1044
	6·000	85·952	4·0267	2·5037	1·6083	0·6062	0·1054
	4·000	87·330	4·0386	2·5078	1·6104	0·6004	0·1060
	2·000	88·674	4·0455	2·5101	1·6116	0·5971	0·1064
1·95	2·000	32·530	1·1162	1·0815	1·0321	1·8787	0·0001
	4·000	34·309	1·2430	1·1673	1·0648	1·8081	0·0004
	6·000	36·199	1·3811	1·2575	1·0983	1·7373	0·0014
	8·000	38·216	1·5315	1·3520	1·1328	1·6658	0·0031
	10·000	40·376	1·6956	1·4508	1·1688	1·5927	0·0060
	12·000	42·709	1·8750	1·5541	1·2065	1·5172	0·0101
	14·000	45·258	2·0724	1·6623	1·2467	1·4381	0·0157
	16·000	48·097	2·2919	1·7767	1·2900	1·3536	0·0230
	18·000	51·373	2·5422	1·8999	1·3381	1·2601	0·0326
	20·000	55·466	2·8455	2·0398	1·3950	1·1492	0·0456
	22·000	63·393	3·3819	2·2650	1·4931	0·9527	0·0714
	22·046	64·699	3·4615	2·2962	1·5075	0·9229	0·0755
	22·000	65·979	3·5367	2·3252	1·5210	0·8945	0·0793
	20·000	72·833	3·8857	2·4541	1·5833	0·7571	0·0977
	18·000	75·902	4·0090	2·4975	1·6052	0·7055	0·1044
	16·000	78·206	4·0870	2·5244	1·6190	0·6718	0·1086
	14·000	80·129	4·1420	2·5431	1·6287	0·6473	0·1116
	12·000	81·820	4·1826	2·5568	1·6359	0·6288	0·1138
	10·000	83·358	4·2131	2·5670	1·6412	0·6147	0·1155
	8·000	84·791	4·2359	2·5746	1·6453	0·6040	0·1168
	6·000	86·150	4·2525	2·5801	1·6482	0·5962	0·1177
	4·000	87·459	4·2638	2·5839	1·6502	0·5908	0·1183
	2·000	88·737	4·2703	2·5860	1·6513	0·5876	0·1187
2·00	2·000	31·648	1·1182	1·0829	1·0327	1·9278	0·0001
	4·000	33·395	1·2473	1·1702	1·0659	1·8564	0·0005
	6·000	35·249	1·3880	1·2620	1·0999	1·7849	0·0015
	8·000	37·221	1·5413	1·3580	1·1350	1·7129	0·0033
	10·000	39·329	1·7084	1·4583	1·1715	1·6395	0·0062
	12·000	41·596	1·8908	1·5629	1·2098	1·5639	0·0105
	14·000	44·055	2·0907	1·6721	1·2503	1·4851	0·0162
	16·000	46·766	2·3116	1·7867	1·2938	1·4017	0·0237
	18·000	49·833	2·5598	1·9083	1·3414	1·3110	0·0333
	20·000	53·493	2·8500	2·0418	1·3958	1·2076	0·0458
	22·000	58·596	3·2349	2·2059	1·4665	1·0719	0·0640
	22·925	64·650	3·6469	2·3670	1·5408	0·9243	0·0851
	22·000	70·182	3·9663	2·4826	1·5976	0·8046	0·1020
	20·000	74·190	4·1564	2·5480	1·6312	0·7292	0·1124
	18·000	76·804	4·2597	2·5825	1·6494	0·6863	0·1181
	16·000	78·877	4·3293	2·6054	1·6617	0·6565	0·1219
	14·000	80·649	4·3798	2·6218	1·6706	0·6343	0·1247
	12·000	82·229	4·4177	2·6340	1·6772	0·6173	0·1268
	10·000	83·678	4·4465	2·6432	1·6823	0·6042	0·1285
	8·000	85·035	4·4681	2·6501	1·6861	0·5942	0·1296
	6·000	86·327	4·4839	2·6551	1·6888	0·5868	0·1305

TABLE V

FLOW OF DRY AIR THROUGH A PLANE OBLIQUE SHOCK WAVE

M_1	δ	β	p_2/p_1	ρ_2/ρ_1	T_2/T_1	M_2	$\Delta S/c_v$
2·00	4·000	87·574	4·4948	2·6585	1·6907	0·5817	0·131
	2·000	88·794	4·5011	2·6605	1·6918	0·5787	0·131
2·05	2·000	30·819	1·1203	1·0843	1·0332	1·9768	0·000
	4·000	32·536	1·2518	1·1732	1·0670	1·9045	0·000
	6·000	34·358	1·3952	1·2666	1·1016	1·8323	0·001
	8·000	36·293	1·5516	1·3643	1·1373	1·7596	0·003
	10·000	38·356	1·7220	1·4662	1·1744	1·6856	0·006
	12·000	40·566	1·9077	1·5724	1·2133	1·6098	0·011
	14·000	42·953	2·1108	1·6828	1·2543	1·5311	0·016
	16·000	45·562	2·3341	1·7980	1·2982	1·4482	0·024
	18·000	48·472	2·5826	1·9191	1·3457	1·3594	0·034
	20·000	51·847	2·8668	2·0493	1·3989	1·2605	0·046
	22·000	56·133	3·2157	2·1980	1·4630	1·1410	0·063
	23·763	64·618	3·8379	2·4370	1·5748	0·9258	0·095
	22·000	72·082	4·2751	2·5876	1·6522	0·7647	0·119
	20·000	75·253	4·4216	2·6352	1·6779	0·7068	0·127
	18·000	77·560	4·5119	2·6639	1·6937	0·6697	0·132
	16·000	79·456	4·5752	2·6837	1·7048	0·6429	0·135
	14·000	81·104	4·6222	2·6983	1·7130	0·6225	0·138
	12·000	82·590	4·6580	2·7093	1·7193	0·6067	0·140
	10·000	83·962	4·6853	2·7176	1·7240	0·5945	0·141
	8·000	85·252	4·7060	2·7239	1·7277	0·5850	0·142
	6·000	86·485	4·7211	2·7285	1·7303	0·5781	0·143
	4·000	87·677	4·7315	2·7316	1·7321	0·5732	0·144
	2·000	88·845	4·7376	2·7335	1·7332	0·5704	0·144
2·10	2·000	30·035	1·1224	1·0858	1·0338	2·0258	0·000
	4·000	31·728	1·2564	1·1763	1·0681	1·9525	0·000
	6·000	33·520	1·4027	1·2713	1·1033	1·8794	0·001
	8·000	35·423	1·5623	1·3708	1·1397	1·8059	0·003
	10·000	37·448	1·7361	1·4745	1·1774	1·7314	0·006
	12·000	39·611	1·9256	1·5823	1·2169	1·6550	0·011
	14·000	41·937	2·1323	1·6942	1·2586	1·5761	0·017
	16·000	44·463	2·3588	1·8104	1·3030	1·4936	0·025
	18·000	47·251	2·6092	1·9317	1·3508	1·4057	0·035
	20·000	50·421	2·8914	2·0601	1·4035	1·3097	0·047
	22·000	54·253	3·2243	2·2016	1·4646	1·1987	0·063
	24·000	59·965	3·6918	2·3837	1·5488	1·0435	0·087
	24·561	64·600	4·0344	2·5063	1·6097	0·9273	0·105
	24·000	68·896	4·3143	2·6005	1·6590	0·8287	0·121
	22·000	73·428	4·5630	2·6799	1·7027	0·7362	0·134
	20·000	76·124	4·6857	2·7177	1·7241	0·6881	0·141
	18·000	78·207	4·7668	2·7423	1·7383	0·6552	0·146
	16·000	79·960	4·8254	2·7598	1·7485	0·6307	0·149
	14·000	81·506	4·8695	2·7728	1·7562	0·6118	0·152
	12·000	82·911	4·9035	2·7828	1·7621	0·5970	0·154
	10·000	84·215	4·9296	2·7904	1·7666	0·5854	0·155
	8·000	85·447	4·9494	2·7962	1·7701	0·5765	0·156
	6·000	86·627	4·9640	2·8004	1·7726	0·5699	0·157
	4·000	87·770	4·9741	2·8033	1·7744	0·5653	0·158
	2·000	88·890	4·9800	2·8050	1·7754	0·5626	0·158

TABLE V

FLOW OF DRY AIR THROUGH A PLANE OBLIQUE SHOCK WAVE

M_1	δ	β	p_2/p_1	ρ_2/ρ_1	T_2/T_1	M_2	$\Delta S/c_v$
2·15	2·000	29·295	1·1246	1·0873	1·0344	2·0746	0·0001
	4·000	30·964	1·2611	1·1794	1·0693	2·0003	0·0005
	6·000	32·732	1·4104	1·2762	1·1051	1·9263	0·0017
	8·000	34·606	1·5733	1·3776	1·1421	1·8520	0·0038
	10·000	36·598	1·7509	1·4831	1·1806	1·7767	0·0072
	12·000	38·720	1·9443	1·5927	1·2208	1·6997	0·0119
	14·000	40·995	2·1551	1·7061	1·2631	1·6204	0·0183
	16·000	43·453	2·3854	1·8236	1·3081	1·5378	0·0264
	18·000	46·144	2·6389	1·9456	1·3564	1·4506	0·0366
	20·000	49·159	2·9216	2·0734	1·4091	1·3564	0·0491
	22·000	52·692	3·2471	2·2109	1·4687	1·2504	0·0647
	24·000	57·347	3·6587	2·3714	1·5429	1·1180	0·0857
	25·321	64·594	4·2365	2·5748	1·6454	0·9289	0·1168
	24·000	71·025	4·6593	2·7097	1·7195	0·7821	0·1403
	22·000	74·480	4·8436	2·7651	1·7516	0·7137	0·1507
	20·000	76·859	4·9510	2·7966	1·7704	0·6720	0·1567
	18·000	78·768	5·0252	2·8180	1·7833	0·6422	0·1609
	16·000	80·405	5·0801	2·8336	1·7928	0·6195	0·1640
	14·000	81·864	5·1219	2·8454	1·8000	0·6019	0·1664
	12·000	83·199	5·1543	2·8545	1·8057	0·5879	0·1682
	10·000	84·443	5·1794	2·8615	1·8100	0·5770	0·1696
	8·000	85·623	5·1986	2·8669	1·8133	0·5685	0·1707
	6·000	86·755	5·2127	2·8708	1·8158	0·5622	0·1715
	4·000	87·854	5·2224	2·8735	1·8175	0·5578	0·1721
	2·000	88·932	5·2281	2·8750	1·8185	0·5553	0·1724
2·20	2·000	28·594	1·1268	1·0888	1·0349	2·1234	0·0001
	4·000	30·243	1·2660	1·1826	1·0705	2·0480	0·0005
	6·000	31·988	1·4183	1·2813	1·1070	1·9730	0·0017
	8·000	33·837	1·5847	1·3845	1·1446	1·8977	0·0040
	10·000	35·799	1·7661	1·4919	1·1838	1·8216	0·0075
	12·000	37·887	1·9637	1·6034	1·2247	1·7439	0·0125
	14·000	40·118	2·1789	1·7186	1·2678	1·6641	0·0191
	16·000	42·519	2·4137	1·8376	1·3135	1·5812	0·0275
	18·000	45·131	2·6710	1·9605	1·3624	1·4941	0·0380
	20·000	48·025	2·9560	2·0884	1·4154	1·4010	0·0507
	22·000	51·345	3·2789	2·2238	1·4745	1·2983	0·0662
	24·000	55·460	3·6672	2·3745	1·5444	1·1767	0·0861
	26·000	63·324	4·3449	2·6104	1·6644	0·9627	0·1228
	26·045	64·597	4·4440	2·6424	1·6818	0·9305	0·1283
	26·000	65·842	4·5374	2·6719	1·6982	0·8998	0·1335
	24·000	72·446	4·9699	2·8021	1·7736	0·7511	0·1578
	22·000	75·344	5·1222	2·8455	1·8001	0·6949	0·1664
	20·000	77·491	5·2188	2·8725	1·8169	0·6578	0·1719
	18·000	79·261	5·2878	2·8914	1·8288	0·6304	0·1758
	16·000	80·801	5·3395	2·9055	1·8378	0·6094	0·1787
	14·000	82·184	5·3795	2·9162	1·8447	0·5927	0·1810
	12·000	83·457	5·4106	2·9246	1·8501	0·5795	0·1827
	10·000	84·649	5·4348	2·9310	1·8542	0·5691	0·1841
	8·000	85·782	5·4534	2·9360	1·8575	0·5610	0·1852
	6·000	86·871	5·4672	2·9396	1·8598	0·5550	0·1859
	4·000	87·930	5·4766	2·9421	1·8615	0·5508	0·1865
	2·000	88·970	5·4822	2·9436	1·8624	0·5483	0·1868

TABLE V

FLOW OF DRY AIR THROUGH A PLANE OBLIQUE SHOCK WAVE

M_1	δ	β	p_2/p_1	ρ_2/ρ_1	T_2/T_1	M_2	$\Delta S/c_1$
2·25	2·000	27·928	1·1291	1·0903	1·0355	2·1722	0·000
	4·000	29·559	1·2709	1·1859	1·0717	2·0956	0·000
	6·000	31·284	1·4264	1·2864	1·1088	2·0195	0·001
	8·000	33·112	1·5964	1·3915	1·1472	1·9432	0·004
	10·000	35·048	1·7818	1·5010	1·1871	1·8661	0·007
	12·000	37·106	1·9838	1·6144	1·2288	1·7876	0·013
	14·000	39·300	2·2038	1·7315	1·2727	1·7071	0·019
	16·000	41·652	2·4434	1·8521	1·3192	1·6238	0·028
	18·000	44·198	2·7053	1·9763	1·3688	1·5366	0·039
	20·000	46·995	2·9937	2·1047	1·4224	1·4441	0·052
	22·000	50·154	3·3169	2·2391	1·4814	1·3434	0·068
	24·000	53·927	3·6943	2·3846	1·5492	1·2282	0·087
	26·000	59·312	4·2040	2·5640	1·6396	1·0731	0·115
	26·735	64·609	4·6569	2·7089	1·7191	0·9322	0·140
	26·000	69·428	5·0139	2·8148	1·7813	0·8157	0·160
	24·000	73·535	5·2689	2·8862	1·8255	0·7273	0·174
	22·000	76·073	5·4014	2·9221	1·8485	0·6788	0·182
	20·000	78·042	5·4900	2·9456	1·8638	0·6451	0·187
	18·000	79·698	5·5548	2·9626	1·8750	0·6198	0·190
	16·000	81·155	5·6040	2·9754	1·8835	0·6000	0·193
	14·000	82·473	5·6424	2·9852	1·8901	0·5843	0·195
	12·000	83·691	5·6724	2·9929	1·8953	0·5717	0·197
	10·000	84·836	5·6959	2·9989	1·8993	0·5618	0·198
	8·000	85·927	5·7140	3·0035	1·9024	0·5540	0·199
	6·000	86·977	5·7274	3·0069	1·9047	0·5482	0·200
	4·000	87·999	5·7366	3·0092	1·9063	0·5442	0·201
	2·000	89·004	5·7420	3·0106	1·9073	0·5418	0·2015
2·30	2·000	27·296	1·1314	1·0919	1·0362	2·2209	0·0001
	4·000	28·910	1·2760	1·1892	1·0730	2·1431	0·0006
	6·000	30·618	1·4346	1·2916	1·1108	2·0658	0·0019
	8·000	32·425	1·6083	1·3987	1·1498	1·9885	0·0044
	10·000	34·340	1·7979	1·5103	1·1905	1·9104	0·0082
	12·000	36·371	2·0046	1·6258	1·2330	1·8309	0·0136
	14·000	38·533	2·2295	1·7448	1·2778	1·7496	0·0208
	16·000	40·844	2·4744	1·8672	1·3252	1·6656	0·0299
	18·000	43·334	2·7414	1·9929	1·3756	1·5782	0·0410
	20·000	46·052	3·0343	2·1221	1·4298	1·4859	0·0543
	22·000	49·085	3·3598	2·2562	1·4891	1·3865	0·0703
	24·000	52·618	3·7325	2·3987	1·5561	1·2752	0·0896
	26·000	57·216	4·1984	2·5621	1·6386	1·1376	0·1147
	27·393	64·628	4·8753	2·7745	1·7572	0·9338	0·1525
	26·000	71·117	5·3624	2·9116	1·8417	0·7773	0·1800
	24·000	74·421	5·5639	2·9650	1·8765	0·7077	0·1914
	22·000	76·703	5·6827	2·9956	1·8970	0·6647	0·1981
	20·000	78·529	5·7652	3·0164	1·9113	0·6337	0·2028
	18·000	80·089	5·8265	3·0317	1·9218	0·6100	0·2063
	16·000	81·473	5·8737	3·0434	1·9300	0·5913	0·2090
	14·000	82·734	5·9106	3·0525	1·9363	0·5764	0·2111
	12·000	83·904	5·9398	3·0596	1·9413	0·5644	0·2127
	10·000	85·006	5·9627	3·0652	1·9453	0·5548	0·2140
	8·000	86·059	5·9803	3·0695	1·9483	0·5474	0·2150
	6·000	87·074	5·9934	3·0727	1·9505	0·5418	0·2157

TABLE V

FLOW OF DRY AIR THROUGH A PLANE OBLIQUE SHOCK WAVE

M_1	δ	β	p_2/p_1	ρ_2/ρ_1	T_2/T_1	M_2	$\Delta S/c_v$
2·30	4·000	88·063	6·0024	3·0748	1·9521	0·5380	0·2162
	2·000	89·036	6·0077	3·0761	1·9530	0·5357	0·2166
2·35	2·000	26·695	1·1337	1·0935	1·0368	2·2695	0·0001
	4·000	28·293	1·2811	1·1926	1·0742	2·1905	0·0006
	6·000	29·985	1·4430	1·2969	1·1127	2·1120	0·0020
	8·000	31·775	1·6205	1·4061	1·1525	2·0335	0·0046
	10·000	33·670	1·8144	1·5197	1·1939	1·9543	0·0086
	12·000	35·679	2·0259	1·6373	1·2373	1·8739	0·0142
	14·000	37·812	2·2561	1·7585	1·2830	1·7916	0·0217
	16·000	40·088	2·5065	1·8828	1·3313	1·7068	0·0311
	18·000	42·531	2·7791	2·0100	1·3826	1·6189	0·0426
	20·000	45·183	3·0773	2·1404	1·4377	1·5265	0·0564
	22·000	48·115	3·4065	2·2747	1·4976	1·4278	0·0727
	24·000	51·468	3·7785	2·4155	1·5642	1·3192	0·0920
	26·000	55·616	4·2243	2·5708	1·6432	1·1909	0·1162
	28·000	63·834	5·0270	2·8185	1·7836	0·9571	0·1610
	28·019	64·653	5·0991	2·8390	1·7961	0·9355	0·1651
	28·000	65·459	5·1685	2·8585	1·8081	0·9144	0·1690
	26·000	72·330	5·6868	2·9966	1·8977	0·7499	0·1984
	24·000	75·167	5·8584	3·0396	1·9273	0·6911	0·2081
	22·000	77·253	5·9670	3·0663	1·9460	0·6521	0·2142
	20·000	78·963	6·0446	3·0850	1·9594	0·6234	0·2186
	18·000	80·441	6·1031	3·0989	1·9694	0·6011	0·2219
	16·000	81·762	6·1486	3·1097	1·9772	0·5833	0·2245
	14·000	82·971	6·1844	3·1181	1·9834	0·5690	0·2265
	12·000	84·098	6·2128	3·1247	1·9883	0·5575	0·2281
	10·000	85·162	6·2351	3·1299	1·9921	0·5484	0·2293
	8·000	86·179	6·2523	3·1339	1·9950	0·5412	0·2303
	6·000	87·162	6·2652	3·1369	1·9972	0·5358	0·2310
	4·000	88·121	6·2740	3·1389	1·9988	0·5321	0·2315
	2·000	89·064	6·2792	3·1401	1·9997	0·5298	0·2318
2·40	2·000	26·121	1·1361	1·0951	1·0374	2·3181	0·0001
	4·000	27·706	1·2863	1·1960	1·0755	2·2377	0·0006
	6·000	29·384	1·4516	1·3023	1·1147	2·1580	0·0021
	8·000	31·159	1·6329	1·4135	1·1552	2·0783	0·0048
	10·000	33·036	1·8313	1·5293	1·1975	1·9979	0·0090
	12·000	35·024	2·0478	1·6491	1·2417	1·9164	0·0149
	14·000	37·134	2·2834	1·7724	1·2883	1·8331	0·0227
	16·000	39·379	2·5396	1·8987	1·3376	1·7475	0·0325
	18·000	41·782	2·8183	2·0276	1·3899	1·6588	0·0444
	20·000	44·378	3·1223	2·1593	1·4460	1·5662	0·0585
	22·000	47·227	3·4565	2·2943	1·5066	1·4678	0·0752
	24·000	50·441	3·8302	2·4343	1·5735	1·3609	0·0948
	26·000	54·287	4·2665	2·5847	1·6506	1·2383	0·1185
	28·000	59·882	4·8648	2·7714	1·7553	1·0705	0·1518
	28·616	64·683	5·3283	2·9024	1·8358	0·9371	0·1781
	28·000	69·056	5·6989	2·9997	1·8998	0·8254	0·1991
	26·000	73·289	6·0021	3·0748	1·9521	0·7282	0·2162
	24·000	75·809	6·1541	3·1110	1·9782	0·6766	0·2248
	22·000	77·741	6·2550	3·1346	1·9955	0·6408	0·2305
	20·000	79·352	6·3286	3·1515	2·0081	0·6139	0·2346

TABLE V

FLOW OF DRY AIR THROUGH A PLANE OBLIQUE SHOCK WAVE

M_1	δ	β	p_2/p_1	ρ_2/ρ_1	T_2/T_1	M_2	$\Delta S/c_v$
2·40	18·000	80·759	6·3848	3·1643	2·0178	0·5928	0·237
	16·000	82·025	6·4288	3·1742	2·0253	0·5758	0·240
	14·000	83·188	6·4637	3·1820	2·0313	0·5621	0·242
	12·000	84·275	6·4914	3·1882	2·0360	0·5511	0·243
	10·000	85·305	6·5132	3·1931	2·0398	0·5422	0·245
	8·000	86·290	6·5301	3·1968	2·0427	0·5353	0·245
	6·000	87·244	6·5427	3·1996	2·0448	0·5301	0·246
	4·000	88·175	6·5515	3·2015	2·0463	0·5265	0·247
	2·000	89·091	6·5566	3·2027	2·0472	0·5243	0·247
2·45	2·000	25·574	1·1384	1·0967	1·0380	2·3666	0·000
	4·000	27·147	1·2915	1·1994	1·0768	2·2849	0·000
	6·000	28·812	1·4602	1·3077	1·1166	2·2039	0·002
	8·000	30·573	1·6456	1·4211	1·1580	2·1229	0·005
	10·000	32·435	1·8486	1·5391	1·2011	2·0413	0·009
	12·000	34·405	2·0701	1·6611	1·2462	1·9586	0·015
	14·000	36·493	2·3114	1·7866	1·2938	1·8742	0·023
	16·000	38·712	2·5738	1·9149	1·3441	1·7876	0·033
	18·000	41·080	2·8589	2·0457	1·3975	1·6982	0·046
	20·000	43·630	3·1693	2·1789	1·4545	1·6050	0·060
	22·000	46·409	3·5092	2·3147	1·5161	1·5066	0·077
	24·000	49·512	3·8866	2·4545	1·5835	1·4007	0·097
	26·000	53·138	4·3191	2·6020	1·6599	1·2819	0·121
	28·000	57·945	4·8671	2·7721	1·7557	1·1325	0·152
	29·186	64·717	5·5629	2·9647	1·8764	0·9387	0·191
	28·000	70·655	6·0723	3·0916	1·9641	0·7874	0·220
	26·000	74·084	6·3143	3·1482	2·0057	0·7101	0·233
	24·000	76·371	6·4523	3·1795	2·0293	0·6638	0·241
	22·000	78·176	6·5471	3·2006	2·0456	0·6306	0·246
	20·000	79·703	6·6175	3·2160	2·0577	0·6052	0·250
	18·000	81·048	6·6717	3·2278	2·0669	0·5851	0·253
	16·000	82·265	6·7145	3·2371	2·0743	0·5689	0·256
	14·000	83·387	6·7485	3·2443	2·0801	0·5557	0·258
	12·000	84·438	6·7756	3·2501	2·0847	0·5450	0·259
	10·000	85·436	6·7971	3·2547	2·0884	0·5365	0·260
	8·000	86·393	6·8137	3·2582	2·0912	0·5298	0·261
	6·000	87·319	6·8261	3·2608	2·0934	0·5247	0·262
	4·000	88·224	6·8347	3·2627	2·0948	0·5212	0·262
	2·000	89·116	6·8398	3·2637	2·0957	0·5191	0·263
2·50	2·000	25·052	1·1408	1·0984	1·0386	2·4151	0·000
	4·000	26·613	1·2968	1·2029	1·0781	2·3320	0·000
	6·000	28·266	1·4690	1·3132	1·1187	2·2496	0·002
	8·000	30·015	1·6585	1·4288	1·1608	2·1673	0·005
	10·000	31·863	1·8661	1·5490	1·2047	2·0844	0·009
	12·000	33·818	2·0930	1·6733	1·2508	2·0005	0·016
	14·000	35·887	2·3401	1·8010	1·2993	1·9149	0·024
	16·000	38·083	2·6088	1·9315	1·3507	1·8272	0·035
	18·000	40·422	2·9007	2·0642	1·4052	1·7368	0·048
	20·000	42·930	3·2180	2·1990	1·4634	1·6430	0·063
	22·000	45·652	3·5644	2·3358	1·5260	1·5444	0·080
	24·000	48·664	3·9468	2·4758	1·5942	1·4391	0·101
	26·000	52·123	4·3794	2·6216	1·6705	1·3227	0·124

TABLE V

FLOW OF DRY AIR THROUGH A PLANE OBLIQUE SHOCK WAVE

M_1	δ	β	p_2/p_1	ρ_2/ρ_1	T_2/T_1	M_2	$\Delta S/c_V$
2·50	28·000	56·472	4·9039	2·7829	1·7622	1·1834	0·1541
	29·729	64·754	5·8028	3·0259	1·9178	0·9402	0·2050
	28·000	71·804	6·4188	3·1720	2·0236	0·7604	0·2397
	26·000	74·762	6·6263	3·2180	2·0592	0·6947	0·2513
	24·000	76·867	6·7537	3·2455	2·0810	0·6523	0·2584
	22·000	78·567	6·8437	3·2646	2·0964	0·6212	0·2634
	20·000	80·022	6·9113	3·2787	2·1079	0·5971	0·2672
	18·000	81·312	6·9639	3·2897	2·1169	0·5779	0·2701
	16·000	82·484	7·0056	3·2982	2·1240	0·5623	0·2724
	14·000	83·570	7·0389	3·3051	2·1297	0·5496	0·2742
	12·000	84·589	7·0656	3·3105	2·1343	0·5393	0·2757
	10·000	85·557	7·0867	3·3148	2·1379	0·5310	0·2769
	8·000	86·487	7·1031	3·3181	2·1407	0·5245	0·2778
	6·000	87·389	7·1153	3·3206	2·1428	0·5196	0·2784
	4·000	88·270	7·1238	3·3223	2·1442	0·5162	0·2789
	2·000	89·138	7·1288	3·3233	2·1451	0·5142	0·2792
2·55	2·000	24·552	1·1432	1·1000	1·0393	2·4636	0·0001
	4·000	26·103	1·3022	1·2064	1·0794	2·3789	0·0008
	6·000	27·746	1·4780	1·3188	1·1207	2·2951	0·0024
	8·000	29·483	1·6716	1·4365	1·1636	2·2115	0·0055
	10·000	31·320	1·8840	1·5591	1·2084	2·1273	0·0103
	12·000	33·261	2·1163	1·6857	1·2554	2·0420	0·0170
	14·000	35·314	2·3695	1·8157	1·3050	1·9552	0·0258
	16·000	37·489	2·6448	1·9483	1·3575	1·8664	0·0369
	18·000	39·801	2·9436	2·0830	1·4132	1·7750	0·0501
	20·000	42·275	3·2682	2·2194	1·4725	1·6803	0·0657
	22·000	44·947	3·6218	2·3575	1·5363	1·5812	0·0837
	24·000	47·883	4·0103	2·4980	1·6054	1·4761	0·1044
	26·000	51·211	4·4454	2·6428	1·6821	1·3614	0·1284
	28·000	55·251	4·9585	2·7988	1·7717	1·2283	0·1571
	30·000	61·822	5·7322	3·0081	1·9056	1·0266	0·2009
	30·248	64·794	6·0481	3·0858	1·9600	0·9418	0·2188
	30·000	67·584	6·3212	3·1498	2·0069	0·8662	0·2342
	28·000	72·715	6·7550	3·2457	2·0812	0·7391	0·2585
	26·000	75·351	6·9397	3·2846	2·1128	0·6811	0·2687
	24·000	77·311	7·0590	3·3092	2·1332	0·6419	0·2753
	22·000	78·921	7·1449	3·3266	2·1478	0·6126	0·2801
	20·000	80·313	7·2103	3·3396	2·1590	0·5896	0·2837
	18·000	81·554	7·2615	3·3498	2·1677	0·5712	0·2865
	16·000	82·687	7·3023	3·3578	2·1747	0·5562	0·2887
	14·000	83·738	7·3350	3·3642	2·1803	0·5439	0·2906
	12·000	84·727	7·3612	3·3694	2·1847	0·5339	0·2920
	10·000	85·669	7·3820	3·3734	2·1883	0·5259	0·2931
	8·000	86·575	7·3982	3·3765	2·1911	0·5196	0·2940
	6·000	87·453	7·4103	3·3789	2·1931	0·5148	0·2946
	4·000	88·312	7·4187	3·3805	2·1946	0·5115	0·2951
	2·000	89·159	7·4237	3·3815	2·1954	0·5095	0·2954
2·60	2·000	24·073	1·1457	1·1017	1·0399	2·5120	0·0001
	4·000	25·615	1·3077	1·2100	1·0807	2·4258	0·0008
	6·000	27·248	1·4870	1·3244	1·1228	2·3406	0·0026
	8·000	28·976	1·6848	1·4444	1·1664	2·2555	0·0058

TABLE V

FLOW OF DRY AIR THROUGH A PLANE OBLIQUE SHOCK WAVE

M_1	δ	β	p_2/p_1	ρ_2/ρ_1	T_2/T_1	M_2	$\Delta S/c_v$
2·60	10·000	30·802	1·9022	1·5693	1·2121	2·1699	0·0108
	12·000	32·731	2·1400	1·6982	1·2601	2·0833	0·0178
	14·000	34·769	2·3994	1·8305	1·3108	1·9951	0·0270
	16·000	36·926	2·6815	1·9654	1·3644	1·9051	0·0384
	18·000	39·216	2·9877	2·1021	1·4213	1·8125	0·0521
	20·000	41·660	3·3200	2·2403	1·4819	1·7169	0·0683
	22·000	44·289	3·6813	2·3798	1·5469	1·6172	0·0869
	24·000	47·162	4·0767	2·5209	1·6172	1·5120	0·1080
	26·000	50·382	4·5163	2·6653	1·6945	1·3984	0·1323
	28·000	54·198	5·0247	2·8179	1·7832	1·2694	0·1609
	30·000	59·577	5·7019	3·0004	1·9004	1·0982	0·1992
	30·742	64·835	6·2987	3·1446	2·0030	0·9433	0·2329
	30·000	69·546	6·7621	3·2472	2·0824	0·8165	0·2589
	28·000	73·473	7·0872	3·3149	2·1380	0·7213	0·2769
	26·000	75·870	7·2556	3·3486	2·1667	0·6690	0·2862
	24·000	77·711	7·3684	3·3708	2·1860	0·6324	0·2924
	22·000	79·244	7·4510	3·3867	2·2001	0·6046	0·2969
	20·000	80·579	7·5145	3·3989	2·2109	0·5827	0·3003
	18·000	81·777	7·5645	3·4083	2·2194	0·5649	0·3031
	16·000	82·873	7·6045	3·4159	2·2262	0·5504	0·3052
	14·000	83·894	7·6367	3·4219	2·2317	0·5386	0·3070
	12·000	84·856	7·6625	3·4267	2·2361	0·5289	0·3084
	10·000	85·773	7·6831	3·4305	2·2396	0·5210	0·3095
	8·000	86·656	7·6991	3·4335	2·2424	0·5149	0·3104
	6·000	87·513	7·7111	3·4357	2·2444	0·5102	0·3110
	4·000	88·352	7·7195	3·4373	2·2458	0·5070	0·3115
	2·000	89·179	7·7244	3·4382	2·2467	0·5051	0·3118
2·65	2·000	23·615	1·1482	1·1034	1·0406	2·5603	0·0001
	4·000	25·148	1·3132	1·2136	1·0820	2·4726	0·0008
	6·000	26·773	1·4962	1·3301	1·1248	2·3859	0·0027
	8·000	28·491	1·6983	1·4523	1·1693	2·2993	0·0061
	10·000	30·308	1·9206	1·5795	1·2159	2·2123	0·0113
	12·000	32·226	2·1641	1·7109	1·2649	2·1242	0·0186
	14·000	34·252	2·4299	1·8455	1·3166	2·0347	0·0282
	16·000	36·393	2·7191	1·9827	1·3714	1·9433	0·0400
	18·000	38·663	3·0329	2·1215	1·4296	1·8496	0·0543
	20·000	41·080	3·3731	2·2615	1·4915	1·7529	0·0710
	22·000	43·673	3·7426	2·4024	1·5578	1·6525	0·0901
	24·000	46·491	4·1457	2·5444	1·6293	1·5470	0·1118
	26·000	49·623	4·5912	2·6887	1·7076	1·4339	0·1365
	28·000	53·266	5·0991	2·8390	1·7961	1·3077	0·1651
	30·000	58·054	5·7368	3·0093	1·9064	1·1507	0·2012
	31·215	64·879	6·5546	3·2022	2·0469	0·9448	0·2473
	30·000	70·798	7·1455	3·3267	2·1479	0·7856	0·2801
	28·000	74·121	7·4186	3·3805	2·1945	0·7061	0·2951
	26·000	76·333	7·5747	3·4103	2·2212	0·6582	0·3036
	24·000	78·073	7·6823	3·4304	2·2395	0·6237	0·3095
	22·000	79·538	7·7622	3·4451	2·2531	0·5973	0·3138
	20·000	80·824	7·8240	3·4564	2·2636	0·5762	0·3171
	18·000	81·982	7·8730	3·4653	2·2720	0·5590	0·3198
	16·000	83·046	7·9123	3·4724	2·2787	0·5450	0·3219
	14·000	84·038	7·9440	3·4780	2·2841	0·5335	0·3236

TABLE V

FLOW OF DRY AIR THROUGH A PLANE OBLIQUE SHOCK WAVE

M_1	δ	β	p_2/p_1	ρ_2/ρ_1	T_2/T_1	M_2	$\Delta S/c_V$
2·65	12·000	84·975	7·9696	3·4826	2·2884	0·5240	0·3250
	10·000	85·870	7·9900	3·4862	2·2919	0·5164	0·3261
	8·000	86·732	8·0059	3·4890	2·2946	0·5104	0·3269
	6·000	87·569	8·0178	3·4911	2·2966	0·5059	0·3276
	4·000	88·389	8·0260	3·4926	2·2980	0·5027	0·3280
	2·000	89·197	8·0309	3·4935	2·2988	0·5009	0·3283
2·70	2·000	23·175	1·1506	1·1051	1·0412	2·6086	0·0001
	4·000	24·700	1·3187	1·2172	1·0834	2·5193	0·0009
	6·000	26·317	1·5054	1·3359	1·1269	2·4310	0·0028
	8·000	28·028	1·7119	1·4604	1·1723	2·3430	0·0063
	10·000	29·836	1·9393	1·5899	1·2198	2·2544	0·0118
	12·000	31·744	2·1887	1·7237	1·2698	2·1649	0·0194
	14·000	33·759	2·4610	1·8607	1·3226	2·0740	0·0293
	16·000	35·887	2·7574	2·0002	1·3786	1·9812	0·0417
	18·000	38·140	3·0790	2·1411	1·4380	1·8862	0·0564
	20·000	40·533	3·4276	2·2830	1·5014	1·7884	0·0737
	22·000	43·094	3·8056	2·4254	1·5691	1·6870	0·0934
	24·000	45·865	4·2171	2·5684	1·6419	1·5810	0·1157
	26·000	48·924	4·6697	2·7129	1·7213	1·4681	0·1409
	28·000	52·430	5·1801	2·8617	1·8101	1·3440	0·1697
	30·000	56·839	5·7979	3·0246	1·9169	1·1955	0·2047
	31·666	64·924	6·8158	3·2587	2·0916	0·9462	0·2619
	30·000	71·754	7·5104	3·3981	2·2102	0·7622	0·3001
	28·000	74·686	7·7511	3·4431	2·2512	0·6928	0·3132
	26·000	76·749	7·8976	3·4697	2·2762	0·6483	0·3211
	24·000	78·402	8·0008	3·4882	2·2937	0·6157	0·3266
	22·000	79·809	8·0784	3·5018	2·3069	0·5904	0·3308
	20·000	81·050	8·1388	3·5124	2·3172	0·5701	0·3341
	18·000	82·172	8·1870	3·5207	2·3254	0·5535	0·3366
	16·000	83·206	8·2257	3·5274	2·3320	0·5399	0·3387
	14·000	84·172	8·2571	3·5328	2·3373	0·5287	0·3404
	12·000	85·086	8·2824	3·5371	2·3416	0·5195	0·3417
	10·000	85·960	8·3026	3·5405	2·3450	0·5121	0·3428
	8·000	86·802	8·3184	3·5432	2·3477	0·5062	0·3436
	6·000	87·621	8·3302	3·5452	2·3497	0·5018	0·3443
	4·000	88·423	8·3384	3·5466	2·3511	0·4987	0·3447
	2·000	89·214	8·3433	3·5474	2·3519	0·4968	0·3450
2·75	2·000	22·752	1·1531	1·1068	1·0419	2·6569	0·0001
	4·000	24·270	1·3243	1·2209	1·0847	2·5660	0·0009
	6·000	25·880	1·5148	1·3417	1·1290	2·4761	0·0029
	8·000	27·584	1·7257	1·4685	1·1752	2·3864	0·0066
	10·000	29·385	1·9584	1·6004	1·2236	2·2964	0·0123
	12·000	31·285	2·2137	1·7366	1·2747	2·2053	0·0202
	14·000	33·290	2·4927	1·8761	1·3286	2·1129	0·0306
	16·000	35·406	2·7964	2·0178	1·3859	2·0187	0·0434
	18·000	37·643	3·1262	2·1609	1·4467	1·9223	0·0587
	20·000	40·016	3·4834	2·3047	1·5114	1·8233	0·0766
	22·000	42·549	3·8704	2·4487	1·5806	1·7209	0·0969
	24·000	45·280	4·2908	2·5928	1·6549	1·6142	0·1198
	26·000	48·275	4·7514	2·7376	1·7356	1·5013	0·1455
	28·000	51·670	5·2664	2·8855	1·8251	1·3784	0·1746

TABLE V

FLOW OF DRY AIR THROUGH A PLANE OBLIQUE SHOCK WAVE

M_1	δ	β	p_2/p_1	ρ_2/ρ_1	T_2/T_1	M_2	$\Delta S/c_v$
2·75	30·000	55·810	5·8745	3·0436	1·9301	1·2356	0·2090
	32·000	63·138	6·8601	3·2680	2·0992	1·0020	0·2643
	32·098	64·970	7·0823	3·3139	2·1371	0·9476	0·2766
	32·000	66·727	7·2845	3·3543	2·1717	0·8971	0·2878
	30·000	72·535	7·8677	3·4643	2·2711	0·7432	0·3195
	28·000	75·187	8·0858	3·5031	2·3082	0·6810	0·3313
	26·000	77·126	8·2247	3·5272	2·3318	0·6393	0·3387
	24·000	78·704	8·3243	3·5442	2·3487	0·6083	0·3440
	22·000	80·058	8·3998	3·5569	2·3616	0·5840	0·3480
	20·000	81·259	8·4591	3·5668	2·3716	0·5644	0·3511
	18·000	82·349	8·5065	3·5746	2·3797	0·5483	0·3536
	16·000	83·355	8·5448	3·5809	2·3862	0·5351	0·3556
	14·000	84·297	8·5759	3·5860	2·3915	0·5241	0·3573
	12·000	85·190	8·6010	3·5901	2·3957	0·5152	0·3586
	10·000	86·044	8·6210	3·5934	2·3991	0·5079	0·3597
	8·000	86·868	8·6367	3·5959	2·4018	0·5022	0·3605
	6·000	87·669	8·6485	3·5978	2·4038	0·4978	0·3610
	4·000	88·455	8·6567	3·5992	2·4052	0·4948	0·3615
	2·000	89·230	8·6615	3·5999	2·4060	0·4930	0·3617
2·80	2·000	22·346	1·1557	1·1085	1·0425	2·7051	0·0001
	4·000	23·858	1·3300	1·2245	1·0861	2·6125	0·0009
	6·000	25·461	1·5242	1·3475	1·1312	2·5210	0·0030
	8·000	27·159	1·7397	1·4766	1·1782	2·4298	0·0069
	10·000	28·953	1·9776	1·6110	1·2276	2·3381	0·0128
	12·000	30·846	2·2390	1·7497	1·2796	2·2455	0·0211
	14·000	32·842	2·5248	1·8916	1·3348	2·1515	0·0319
	16·000	34·947	2·8362	2·0357	1·3933	2·0558	0·0452
	18·000	37·171	3·1742	2·1809	1·4554	1·9580	0·0610
	20·000	39·526	3·5404	2·3266	1·5217	1·8577	0·0795
	22·000	42·034	3·9367	2·4722	1·5924	1·7542	0·1005
	24·000	44·730	4·3665	2·6175	1·6682	1·6467	0·1240
	26·000	47·671	4·8359	2·7629	1·7503	1·5335	0·1503
	28·000	50·973	5·3572	2·9102	1·8408	1·4114	0·1797
	30·000	54·910	5·9618	3·0650	1·9451	1·2725	0·2139
	32·000	60·722	6·7976	3·2548	2·0885	1·0805	0·2608
	32·511	65·017	7·3540	3·3680	2·1835	0·9490	0·2916
	32·000	68·915	7·8023	3·4525	2·2599	0·8381	0·3160
	30·000	73·196	8·2220	3·5267	2·3313	0·7272	0·3385
	28·000	75·633	8·4235	3·5609	2·3656	0·6703	0·3492
	26·000	77·469	8·5562	3·5828	2·3881	0·6311	0·3562
	24·000	78·981	8·6527	3·5985	2·4045	0·6014	0·3613
	22·000	80·288	8·7266	3·6104	2·4171	0·5780	0·3652
	20·000	81·453	8·7849	3·6197	2·4270	0·5590	0·3682
	18·000	82·513	8·8317	3·6271	2·4349	0·5434	0·3707
	16·000	83·494	8·8696	3·6331	2·4413	0·5305	0·3727
	14·000	84·414	8·9004	3·6379	2·4466	0·5198	0·3742
	12·000	85·287	8·9253	3·6418	2·4508	0·5110	0·3755
	10·000	86·122	8·9452	3·6449	2·4542	0·5040	0·3766
	8·000	86·929	8·9608	3·6473	2·4568	0·4983	0·3774
	6·000	87·715	8·9726	3·6492	2·4588	0·4941	0·3780
	4·000	88·485	8·9807	3·6504	2·4602	0·4911	0·3784
	2·000	89·245	8·9855	3·6512	2·4610	0·4894	0·3787

TABLE V

FLOW OF DRY AIR THROUGH A PLANE OBLIQUE SHOCK WAVE

M_1	δ	β	p_2/p_1	ρ_2/ρ_1	T_2/T_1	M_2	$\Delta S/c_v$
2.85	2.000	21.956	1.1582	1.1103	1.0432	2.7533	0.0001
	4.000	23.461	1.3357	1.2282	1.0875	2.6590	0.0010
	6.000	25.059	1.5338	1.3534	1.1333	2.5658	0.0032
	8.000	26.751	1.7539	1.4848	1.1812	2.4729	0.0072
	10.000	28.539	1.9972	1.6217	1.2315	2.3796	0.0134
	12.000	30.425	2.2647	1.7629	1.2847	2.2854	0.0220
	14.000	32.414	2.5575	1.9072	1.3410	2.1898	0.0332
	16.000	34.510	2.8767	2.0536	1.4008	2.0925	0.0470
	18.000	36.722	3.2233	2.2011	1.4644	1.9932	0.0635
	20.000	39.061	3.5986	2.3488	1.5321	1.8915	0.0825
	22.000	41.548	4.0045	2.4960	1.6044	1.7869	0.1041
	24.000	44.212	4.4442	2.6425	1.6819	1.6785	0.1283
	26.000	47.107	4.9231	2.7885	1.7655	1.5649	0.1552
	28.000	50.331	5.4520	2.9356	1.8572	1.4432	0.1851
	30.000	54.108	6.0572	3.0880	1.9615	1.3070	0.2193
	32.000	59.258	6.8386	3.2635	2.0955	1.1321	0.2631
	32.905	65.064	7.6310	3.4208	2.2307	0.9503	0.3067
	32.000	70.162	8.2247	3.5272	2.3318	0.8055	0.3387
	30.000	73.770	8.5761	3.5861	2.3915	0.7133	0.3573
	28.000	76.036	8.7647	3.6165	2.4235	0.6607	0.3672
	26.000	77.783	8.8923	3.6367	2.4452	0.6234	0.3738
	24.000	79.237	8.9862	3.6513	2.4611	0.5950	0.3787
	22.000	80.502	9.0587	3.6624	2.4734	0.5723	0.3824
	20.000	81.633	9.1162	3.6712	2.4832	0.5539	0.3854
	18.000	82.666	9.1624	3.6782	2.4910	0.5387	0.3878
	16.000	83.623	9.2000	3.6839	2.4974	0.5261	0.3897
	14.000	84.523	9.2306	3.6884	2.5026	0.5157	0.3913
	12.000	85.377	9.2554	3.6922	2.5068	0.5071	0.3926
	10.000	86.196	9.2752	3.6951	2.5101	0.5002	0.3936
	8.000	86.987	9.2908	3.6974	2.5128	0.4947	0.3944
	6.000	87.757	9.3025	3.6992	2.5148	0.4905	0.3950
	4.000	88.513	9.3106	3.7004	2.5161	0.4876	0.3955
	2.000	89.259	9.3154	3.7011	2.5169	0.4859	0.3957
2.90	2.000	21.580	1.1607	1.1120	1.0438	2.8015	0.0001
	4.000	23.080	1.3414	1.2320	1.0888	2.7054	0.0010
	6.000	24.672	1.5434	1.3593	1.1355	2.6105	0.0033
	8.000	26.359	1.7682	1.4931	1.1842	2.5159	0.0075
	10.000	28.142	2.0169	1.6325	1.2355	2.4209	0.0140
	12.000	30.023	2.2908	1.7761	1.2898	2.3250	0.0239
	14.000	32.005	2.5908	1.9230	1.3473	2.2278	0.0346
	16.000	34.093	2.9179	2.0718	1.4084	2.1289	0.0489
	18.000	36.294	3.2732	2.2214	1.4734	2.0281	0.0659
	20.000	38.619	3.6579	2.3711	1.5427	1.9250	0.0856
	22.000	41.087	4.0739	2.5199	1.6167	1.8191	0.1079
	24.000	43.724	4.5239	2.6677	1.6958	1.7097	0.1328
	26.000	46.578	5.0129	2.8145	1.7811	1.5954	0.1602
	28.000	49.736	5.5505	2.9615	1.8742	1.4739	0.1907
	30.000	53.384	6.1590	3.1122	1.9790	1.3396	0.2251
	32.000	58.118	6.9131	3.2791	2.1082	1.1749	0.2673
	33.283	65.111	7.9132	3.4725	2.2788	0.9516	0.3220
	32.000	71.093	8.6216	3.5935	2.3992	0.7817	0.3597
	30.000	74.276	8.9314	3.6428	2.4518	0.7010	0.3759

TABLE V

FLOW OF DRY AIR THROUGH A PLANE OBLIQUE SHOCK WAVE

M_1	δ	β	p_2/p_1	ρ_2/ρ_1	T_2/T_1	M_2	$\Delta S/c_v$
2·90	28·000	76·402	9·1099	3·6702	2·4821	0·6519	0·3851
	26·000	78·071	9·2332	3·6888	2·5030	0·6164	0·3914
	24·000	79·474	9·3250	3·7025	2·5186	0·5890	0·3962
	22·000	80·700	9·3963	3·7130	2·5307	0·5670	0·3998
	20·000	81·801	9·4530	3·7213	2·5403	0·5491	0·4027
	18·000	82·808	9·4988	3·7279	2·5480	0·5343	0·4050
	16·000	83·745	9·5361	3·7333	2·5544	0·5220	0·4069
	14·000	84·625	9·5665	3·7377	2·5595	0·5118	0·4085
	12·000	85·462	9·5912	3·7412	2·5637	0·5034	0·4097
	10·000	86·265	9·6110	3·7440	2·5670	0·4966	0·4107
	8·000	87·041	9·6265	3·7462	2·5697	0·4912	0·4115
	6·000	87·797	9·6382	3·7479	2·5716	0·4871	0·4121
	4·000	88·539	9·6463	3·7490	2·5730	0·4843	0·4125
	2·000	89·272	9·6511	3·7497	2·5738	0·4826	0·4128
2·95	2·000	21·218	1·1633	1·1137	1·0445	2·8496	0·0001
	4·000	22·712	1·3472	1·2357	1·0902	2·7517	0·0011
	6·000	24·300	1·5532	1·3653	1·1376	2·6550	0·0035
	8·000	25·983	1·7827	1·5015	1·1873	2·5587	0·0079
	10·000	27·761	2·0370	1·6433	1·2396	2·4620	0·0146
	12·000	29·637	2·3172	1·7895	1·2949	2·3644	0·0239
	14·000	31·613	2·6245	1·9388	1·3536	2·2655	0·0360
	16·000	33·694	2·9597	2·0900	1·4161	2·1650	0·0508
	18·000	35·886	3·3239	2·2419	1·4826	2·0625	0·0685
	20·000	38·199	3·7184	2·3935	1·5535	1·9579	0·0888
	22·000	40·649	4·1446	2·5440	1·6292	1·8507	0·1118
	24·000	43·263	4·6053	2·6931	1·7101	1·7402	0·1373
	26·000	46·081	5·1049	2·8407	1·7971	1·6252	0·1654
	28·000	49·181	5·6522	2·9878	1·8918	1·5036	0·1964
	30·000	52·722	6·2664	3·1372	1·9975	1·3706	0·2311
	32·000	57·163	7·0063	3·2984	2·1242	1·2126	0·2724
	33·645	65·158	8·2007	3·5231	2·3277	0·9528	0·3374
	32·000	71·848	9·0080	3·6546	2·4648	0·7625	0·3798
	30·000	74·727	9·2891	3·6972	2·5125	0·6901	0·3943
	28·000	76·736	9·4593	3·7222	2·5413	0·6437	0·4030
	26·000	78·338	9·5790	3·7394	2·5616	0·6098	0·4091
	24·000	79·694	9·6690	3·7522	2·5769	0·5833	0·4136
	22·000	80·885	9·7393	3·7621	2·5888	0·5621	0·4172
	20·000	81·958	9·7954	3·7700	2·5983	0·5446	0·4201
	18·000	82·942	9·8408	3·7763	2·6060	0·5302	0·4223
	16·000	83·858	9·8779	3·7814	2·6122	0·5181	0·4243
	14·000	84·721	9·9082	3·7856	2·6174	0·5081	0·4257
	12·000	85·542	9·9329	3·7889	2·6215	0·4999	0·4269
	10·000	86·330	9·9526	3·7916	2·6249	0·4932	0·4279
	8·000	87·092	9·9681	3·7937	2·6275	0·4879	0·4287
	6·000	87·835	9·9798	3·7953	2·6295	0·4839	0·4293
	4·000	88·564	9·9879	3·7964	2·6309	0·4811	0·4297
	2·000	89·284	9·9927	3·7971	2·6317	0·4794	0·4299
3·00	2·000	20·869	1·1659	1·1155	1·0452	2·8976	0·0002
	4·000	22·358	1·3530	1·2395	1·0916	2·7979	0·0011
	6·000	23·942	1·5630	1·3713	1·1398	2·6995	0·0036
	8·000	25·620	1·7973	1·5099	1·1903	2·6014	0·0082

TABLE V

FLOW OF DRY AIR THROUGH A PLANE OBLIQUE SHOCK WAVE

M_1	δ	β	p_2/p_1	ρ_2/ρ_1	T_2/T_1	M_2	$\Delta S/c_v$
3·00	10·000	27·395	2·0572	1·6542	1·2436	2·5029	0·0152
	12·000	29·267	2·3440	1·8030	1·3001	2·4036	0·0249
	14·000	31·238	2·6587	1·9548	1·3601	2·3029	0·0374
	16·000	33·313	3·0022	2·1084	1·4239	2·2007	0·0528
	18·000	35·496	3·3756	2·2625	1·4920	2·0966	0·0711
	20·000	37·798	3·7799	2·4160	1·5645	1·9905	0·0921
	22·000	40·234	4·2168	2·5683	1·6419	1·8818	0·1157
	24·000	42·826	4·6885	2·7186	1·7246	1·7701	0·1419
	26·000	45·612	5·1993	2·8670	1·8135	1·6543	0·1708
	28·000	48·663	5·7569	3·0144	1·9098	1·5324	0·2023
	30·000	52·114	6·3785	3·1629	2·0167	1·4003	0·2374
	32·000	56·333	7·1120	3·3199	2·1422	1·2471	0·2783
	33·992	65·205	8·4933	3·5725	2·3774	0·9541	0·3529
	32·000	72·485	9·3898	3·7120	2·5296	0·7464	0·3995
	30·000	75·133	9·6498	3·7495	2·5736	0·6801	0·4127
	28·000	77·043	9·8133	3·7725	2·6013	0·6363	0·4209
	26·000	78·584	9·9299	3·7885	2·6211	0·6036	0·4268
	24·000	79·898	10·0184	3·8006	2·6360	0·5780	0·4312
	22·000	81·058	10·0877	3·8099	2·6478	0·5574	0·4346
	20·000	82·105	10·1434	3·8173	2·6572	0·5403	0·4374
	18·000	83·068	10·1886	3·8233	2·6648	0·5262	0·4397
	16·000	83·965	10·2255	3·8282	2·6711	0·5144	0·4415
	14·000	84·811	10·2557	3·8322	2·6762	0·5046	0·4430
	12·000	85·617	10·2803	3·8354	2·6803	0·4965	0·4442
	10·000	86·391	10·3000	3·8380	2·6837	0·4900	0·4452
	8·000	87·140	10·3155	3·8400	2·6863	0·4848	0·4459
	6·000	87·870	10·3271	3·8415	2·6883	0·4808	0·4465
	4·000	88·588	10·3353	3·8426	2·6897	0·4780	0·4469
	2·000	89·296	10·3401	3·8432	2·6905	0·4764	0·4472
3·05	2·000	20·532	1·1685	1·1173	1·0459	2·9457	0·0002
	4·000	22·017	1·3589	1·2433	1·0930	2·8441	0·0012
	6·000	23·597	1·5729	1·3773	1·1420	2·7438	0·0038
	8·000	25·272	1·8121	1·5184	1·1934	2·6439	0·0085
	10·000	27·043	2·0777	1·6652	1·2477	2·5436	0·0158
	12·000	28·911	2·3712	1·8165	1·3053	2·4425	0·0259
	14·000	30·878	2·6934	1·9709	1·3666	2·3400	0·0389
	16·000	32·947	3·0454	2·1268	1·4319	2·2360	0·0549
	18·000	35·124	3·4281	2·2832	1·5015	2·1303	0·0738
	20·000	37·416	3·8426	2·4387	1·5757	2·0226	0·0954
	22·000	39·838	4·2903	2·5926	1·6548	1·9125	0·1198
	24·000	42·411	4·7734	2·7443	1·7394	1·7995	0·1467
	26·000	45·170	5·2958	2·8936	1·8302	1·6828	0·1762
	28·000	48·176	5·8646	3·0412	1·9284	1·5604	0·2085
	30·000	51·551	6·4949	3·1890	2·0366	1·4288	0·2440
	32·000	55·597	7·2271	3·3430	2·1619	1·2790	0·2846
	34·000	61·888	8·2831	3·5372	2·3417	1·0626	0·3417
	34·324	65·252	8·7912	3·6207	2·4280	0·9552	0·3686
	34·000	68·353	9·2167	3·6864	2·5002	0·8619	0·3906
	32·000	73·037	9·7702	3·7665	2·5940	0·7325	0·4188
	30·000	75·501	10·0140	3·8000	2·6353	0·6711	0·4310
	28·000	77·325	10·1719	3·8211	2·6620	0·6293	0·4388
	26·000	78·813	10·2860	3·8362	2·6813	0·5979	0·4445

TABLE V

FLOW OF DRY AIR THROUGH A PLANE OBLIQUE SHOCK WAVE

M_1	δ	β	p_2/p_1	ρ_2/ρ_1	T_2/T_1	M_2	$\Delta S/c_v$
3·05	24·000	80·089	10·3731	3·8475	2·6961	0·5731	0·4487
	22·000	81·219	10·4418	3·8564	2·7077	0·5529	0·4521
	20·000	82·243	10·4970	3·8634	2·7170	0·5363	0·4549
	18·000	83·185	10·5419	3·8692	2·7246	0·5224	0·4570
	16·000	84·065	10·5788	3·8738	2·7308	0·5109	0·4588
	14·000	84·896	10·6089	3·8776	2·7359	0·5013	0·4603
	12·000	85·687	10·6335	3·8807	2·7401	0·4933	0·4615
	10·000	86·448	10·6532	3·8832	2·7434	0·4869	0·4624
	8·000	87·185	10·6687	3·8851	2·7469	0·4817	0·4632
	6·000	87·904	10·6804	3·8866	2·7480	0·4778	0·4638
	4·000	88·610	10·6885	3·8876	2·7494	0·4751	0·4642
	2·000	89·307	10·6933	3·8882	2·7502	0·4735	0·4644
3·10	2·000	20·207	1·1711	1·1190	1·0465	2·9937	0·0002
	4·000	21·688	1·3648	1·2471	1·0944	2·8901	0·0012
	6·000	23·264	1·5829	1·3834	1·1442	2·7880	0·0040
	8·000	24·936	1·8270	1·5269	1·1966	2·6863	0·0089
	10·000	26·704	2·0985	1·6763	1·2519	2·5842	0·0164
	12·000	28·569	2·3986	1·8301	1·3106	2·4811	0·0269
	14·000	30·533	2·7286	1·9870	1·3732	2·3769	0·0404
	16·000	32·597	3·0892	2·1454	1·4399	2·2711	0·0570
	18·000	34·768	3·4814	2·3039	1·5111	2·1636	0·0764
	20·000	37·052	3·9063	2·4615	1·5870	2·0543	0·0988
	22·000	39·461	4·3651	2·6170	1·6680	1·9427	0·1239
	24·000	42·017	4·8600	2·7700	1·7545	1·8284	0·1515
	26·000	44·751	5·3944	2·9202	1·8472	1·7106	0·1818
	28·000	47·719	5·9749	3·0682	1·9474	1·5876	0·2147
	30·000	51·028	6·6151	3·2155	2·0572	1·4563	0·2507
	32·000	54·932	7·3497	3·3671	2·1828	1·3090	0·2913
	34·000	60·485	8·3304	3·5452	2·3498	1·1131	0·3443
	34·643	65·298	9·0942	3·6679	2·4794	0·9564	0·3843
	34·000	69·587	9·6888	3·7550	2·5802	0·8277	0·4147
	32·000	73·523	10·1513	3·8184	2·6585	0·7203	0·4378
	30·000	75·838	10·3822	3·8487	2·6976	0·6628	0·4492
	28·000	77·586	10·5354	3·8683	2·7235	0·6229	0·4567
	26·000	79·026	10·6473	3·8824	2·7424	0·5925	0·4622.
	24·000	80·268	10·7333	3·8931	2·7570	0·5683	0·4664
	22·000	81·371	10·8013	3·9015	2·7685	0·5487	0·4697
	20·000	82·372	10·8562	3·9083	2·7778	0·5324	0·4723
	18·000	83·296	10·9010	3·9137	2·7853	0·5188	0·4745
	16·000	84·160	10·9378	3·9182	2·7915	0·5075	0·4763
	14·000	84·975	10·9679	3·9218	2·7966	0·4981	0·4777
	12·000	85·754	10·9924	3·9248	2·8008	0·4902	0·4789
	10·000	86·502	11·0122	3·9272	2·8041	0·4839	0·4798
	8·000	87·228	11·0277	3·9290	2·8067	0·4788	0·4806
	6·000	87·935	11·0394	3·9304	2·8087	0·4750	0·4811
	4·000	88·630	11·0475	3·9314	2·8101	0·4723	0·4815
	2·000	89·317	11·0524	3·9320	2·8109	0·4707	0·4817
3·15	2·000	19·892	1·1737	1·1208	1·0472	3·0416	0·0002
	4·000	21·370	1·3708	1·2509	1·0958	2·9361	0·0013
	6·000	22·943	1·5930	1·3895	1·1465	2·8321	0·0041
	8·000	24·613	1·8421	1·5354	1·1997	2·7285	0·0093

TABLE V

FLOW OF DRY AIR THROUGH A PLANE OBLIQUE SHOCK WAVE

M_1	δ	β	p_2/p_1	ρ_2/ρ_1	T_2/T_1	M_2	$\Delta S/c_v$
3·15	10·000	26·378	2·1195	1·6874	1·2561	2·6245	0·0171
	12·000	28·241	2·4265	1·8439	1·3160	2·5196	0·0280
	14·000	30·201	2·7642	2·0033	1·3799	2·4135	0·0420
	16·000	32·262	3·1336	2·1640	1·4480	2·3058	0·0591
	18·000	34·426	3·5356	2·3248	1·5208	2·1966	0·0792
	20·000	36·703	3·9710	2·4843	1·5985	2·0855	0·1023
	22·000	39·102	4·4413	2·6415	1·6813	1·9724	0·1282
	24·000	41·643	4·9482	2·7958	1·7699	1·8568	0·1566
	26·000	44·354	5·4950	2·9470	1·8646	1·7379	0·1875
	28·000	47·288	6·0878	3·0953	1·9668	1·6142	0·2210
	30·000	50·540	6·7389	3·2423	2·0784	1·4829	0·2576
	32·000	54·327	7·4786	3·3920	2·2048	1·3374	0·2984
	34·000	59·428	8·4215	3·5605	2·3652	1·1535	0·3491
	34·948	65·344	9·4025	3·7139	2·5317	0·9575	0·4001
	34·000	70·481	10·1254	3·8150	2·6541	0·8035	0·4366
	32·000	73·958	10·5340	3·8681	2·7233	0·7093	0·4567
	30·000	76·146	10·7546	3·8958	2·7606	0·6552	0·4674
	28·000	77·829	10·9038	3·9141	2·7858	0·6168	0·4746
	26·000	79·225	11·0139	3·9274	2·8044	0·5874	0·4799
	24·000	80·435	11·0989	3·9375	2·8188	0·5639	0·4840
	22·000	81·513	11·1665	3·9455	2·8302	0·5447	0·4872
	20·000	82·494	11·2211	3·9519	2·8394	0·5287	0·4898
	18·000	83·400	11·2658	3·9571	2·8470	0·5154	0·4919
	16·000	84·249	11·3025	3·9614	2·8532	0·5043	0·4936
	14·000	85·051	11·3326	3·9649	2·8583	0·4950	0·4951
	12·000	85·817	11·3572	3·9677	2·8624	0·4873	0·4963
	10·000	86·554	11·3770	3·9700	2·8658	0·4810	0·4972
	8·000	87·268	11·3925	3·9718	2·8684	0·4761	0·4979
	6·000	87·965	11·4043	3·9731	2·8704	0·4723	0·4985
	4·000	88·650	11·4124	3·9740	2·8717	0·4697	0·4989
	2·000	89·327	11·4173	3·9746	2·8726	0·4681	0·4991
3·20	2·000	19·589	1·1763	1·1226	1·0479	3·0895	0·0002
	4·000	21·063	1·3768	1·2548	1·0972	2·9821	0·0013
	6·000	22·634	1·6032	1·3956	1·1487	2·8761	0·0043
	8·000	24·301	1·8573	1·5441	1·2029	2·7706	0·0096
	10·000	26·064	2·1407	1·6986	1·2603	2·6646	0·0178
	12·000	27·924	2·4547	1·8576	1·3214	2·5578	0·0291
	14·000	29·882	2·8003	2·0196	1·3866	2·4498	0·0436
	16·000	31·939	3·1787	2·1828	1·4563	2·3403	0·0613
	18·000	34·099	3·5905	2·3457	1·5307	2·2292	0·0821
	20·000	36·369	4·0368	2·5072	1·6101	2·1164	0·1059
	22·000	38·759	4·5187	2·6660	1·6949	2·0017	0·1325
	24·000	41·286	5·0380	2·8217	1·7855	1·8847	0·1616
	26·000	43·978	5·5976	2·9737	1·8824	1·7646	0·1934
	28·000	46·881	6·2032	3·1225	1·9866	1·6401	0·2276
	30·000	50·082	6·8661	3·2693	2·1002	1·5086	0·2647
	32·000	53·771	7·6130	3·4174	2·2277	1·3645	0·3057
	34·000	58·555	8·5354	3·5794	2·3846	1·1885	0·3551
	35·242	65·389	9·7159	3·7588	2·5848	0·9586	0·4160
	34·000	71·199	10·5476	3·8699	2·7256	0·7843	0·4573
	32·000	74·349	10·9194	3·9160	2·7884	0·6995	0·4753
	30·000	76·431	11·1314	3·9414	2·8243	0·6482	0·4855

TABLE V

FLOW OF DRY AIR THROUGH A PLANE OBLIQUE SHOCK WAVE

M_1	δ	β	p_2/p_1	ρ_2/ρ_1	T_2/T_1	M_2	$\Delta S/c_V$
3.20	28.000	78.054	11.2773	3.9585	2.8489	0.6112	0.4925
	26.000	79.411	11.3858	3.9710	2.8672	0.5826	0.4976
	24.000	80.592	11.4701	3.9806	2.8815	0.5597	0.5016
	22.000	81.647	11.5373	3.9882	2.8928	0.5409	0.5047
	20.000	82.609	11.5917	3.9944	2.9020	0.5253	0.5073
	18.000	83.499	11.6363	3.9993	2.9096	0.5122	0.5094
	16.000	84.333	11.6730	4.0034	2.9158	0.5012	0.5111
	14.000	85.122	11.7032	4.0068	2.9208	0.4921	0.5125
	12.000	85.876	11.7278	4.0095	2.9250	0.4845	0.5136
	10.000	86.602	11.7476	4.0117	2.9284	0.4783	0.5146
	8.000	87.306	11.7632	4.0134	2.9310	0.4734	0.5153
	6.000	87.993	11.7750	4.0147	2.9330	0.4697	0.5158
	4.000	88.669	11.7832	4.0156	2.9344	0.4671	0.5163
	2.000	89.336	11.7880	4.0161	2.9352	0.4656	0.5164
3.25	2.000	19.295	1.1790	1.1244	1.0486	3.1374	0.0002
	4.000	20.766	1.3828	1.2586	1.0987	3.0279	0.0014
	6.000	22.334	1.6134	1.4018	1.1509	2.9200	0.0045
	8.000	23.999	1.8726	1.5527	1.2060	2.8125	0.0100
	10.000	25.761	2.1621	1.7098	1.2645	2.7046	0.0185
	12.000	27.620	2.4832	1.8715	1.3268	2.5958	0.0302
	14.000	29.575	2.8369	2.0360	1.3934	2.4858	0.0452
	16.000	31.630	3.2244	2.2016	1.4646	2.3744	0.0635
	18.000	33.786	3.6463	2.3667	1.5407	2.2615	0.0850
	20.000	36.049	4.1037	2.5301	1.6219	2.1469	0.1095
	22.000	38.431	4.5975	2.6906	1.7087	2.0306	0.1369
	24.000	40.946	5.1294	2.8475	1.8013	1.9121	0.1668
	26.000	43.619	5.7022	3.0005	1.9004	1.7907	0.1992
	28.000	46.495	6.3210	3.1497	2.0068	1.6654	0.2342
	30.000	49.652	6.9964	3.2963	2.1225	1.5336	0.2719
	32.000	53.257	7.7522	3.4433	2.2514	1.3904	0.3133
	34.000	57.801	8.6642	3.6004	2.4065	1.2201	0.3619
	35.523	65.434	10.0344	3.8027	2.6387	0.9596	0.4320
	34.000	71.802	10.9632	3.9213	2.7958	0.7683	0.4775
	32.000	74.705	11.3080	3.9620	2.8541	0.6906	0.4939
	30.000	76.694	11.5129	3.9855	2.8887	0.6416	0.5036
	28.000	78.264	11.6560	4.0015	2.9129	0.6059	0.5103
	26.000	79.585	11.7631	4.0134	2.9310	0.5781	0.5153
	24.000	80.739	11.8468	4.0225	2.9451	0.5557	0.5192
	22.000	81.773	11.9136	4.0298	2.9564	0.5373	0.5223
	20.000	82.717	11.9680	4.0357	2.9656	0.5219	0.5248
	18.000	83.592	12.0126	4.0404	2.9731	0.5091	0.5269
	16.000	84.412	12.0493	4.0444	2.9793	0.4983	0.5286
	14.000	85.189	12.0795	4.0476	2.9844	0.4893	0.5300
	12.000	85.932	12.1041	4.0502	2.9885	0.4818	0.5311
	10.000	86.648	12.1240	4.0523	2.9919	0.4757	0.5320
	8.000	87.342	12.1397	4.0539	2.9945	0.4709	0.5327
	6.000	88.020	12.1515	4.0552	2.9965	0.4672	0.5332
	4.000	88.686	12.1597	4.0560	2.9979	0.4647	0.5336
	2.000	89.345	12.1646	4.0566	2.9987	0.4631	0.5339
3.30	2.000	19.011	1.1816	1.1262	1.0493	3.1853	0.0002
	4.000	20.479	1.3889	1.2625	1.1001	3.0737	0.0015

TABLE V

FLOW OF DRY AIR THROUGH A PLANE OBLIQUE SHOCK WAVE

M_1	δ	β	p_2/p_1	ρ_2/ρ_1	T_2/T_1	M_2	$\Delta S/c_V$
3·30	6·000	22·045	1·6237	1·4080	1·1532	2·9638	0·0047
	8·000	23·708	1·8882	1·5614	1·2092	2·8543	0·0104
	10·000	25·469	2·1838	1·7211	1·2688	2·7444	0·0192
	12·000	27·326	2·5120	1·8854	1·3323	2·6336	0·0314
	14·000	29·280	2·8740	2·0524	1·4003	2·5216	0·0469
	16·000	31·332	3·2707	2·2204	1·4730	2·4082	0·0658
	18·000	33·484	3·7029	2·3878	1·5508	2·2934	0·0880
	20·000	35·743	4·1715	2·5531	1·6339	2·1771	0·1132
	22·000	38·117	4·6775	2·7152	1·7227	2·0590	0·1413
	24·000	40·621	5·2223	2·8734	1·8174	1·9390	0·1721
	26·000	43·278	5·8086	3·0273	1·9187	1·8164	0·2053
	28·000	46·130	6·4412	3·1770	2·0274	1·6901	0·2409
	30·000	49·247	7·1297	3·3235	2·1452	1·5579	0·2792
	32·000	52·779	7·8958	3·4694	2·2758	1·4152	0·3210
	34·000	57·135	8·8039	3·6227	2·4302	1·2492	0·3692
	35·794	65·478	10·3581	3·8456	2·6935	0·9606	0·4480
	34·000	72·324	11·3762	3·9699	2·8656	0·7545	0·4971
	32·000	75·031	11·7002	4·0064	2·9203	0·6824	0·5124
	30·000	76·939	11·8992	4·0282	2·9539	0·6355	0·5216
	28·000	78·461	12·0398	4·0434	2·9777	0·6009	0·5281
	26·000	79·749	12·1459	4·0546	2·9956	0·5738	0·5330
	24·000	80·878	12·2290	4·0633	3·0096	0·5519	0·5368
	22·000	81·892	12·2957	4·0702	3·0209	0·5338	0·5398
	20·000	82·819	12·3499	4·0759	3·0300	0·5188	0·5423
	18·000	83·680	12·3945	4·0804	3·0375	0·5061	0·5444
	16·000	84·488	12·4313	4·0842	3·0437	0·4955	0·5460
	14·000	85·253	12·4615	4·0873	3·0489	0·4866	0·5474
	12·000	85·986	12·4863	4·0898	3·0530	0·4793	0·5485
	10·000	86·691	12·5063	4·0918	3·0564	0·4733	0·5494
	8·000	87·376	12·5220	4·0934	3·0591	0·4685	0·5501
	6·000	88·046	12·5338	4·0946	3·0611	0·4649	0·5507
	4·000	88·703	12·5421	4·0955	3·0624	0·4623	0·5510
	2·000	89·353	12·5470	4·0959	3·0633	0·4608	0·5513
3·35	2·000	18·736	1·1843	1·1280	1·0499	3·2331	0·0002
	4·000	20·201	1·3950	1·2664	1·1015	3·1194	0·0015
	6·000	21·765	1·6341	1·4143	1·1555	3·0074	0·0048
	8·000	23·427	1·9038	1·5702	1·2125	2·8959	0·0108
	10·000	25·187	2·2057	1·7325	1·2731	2·7840	0·0200
	12·000	27·043	2·5412	1·8994	1·3379	2·6711	0·0325
	14·000	28·995	2·9115	2·0690	1·4072	2·5571	0·0486
	16·000	31·045	3·3176	2·2394	1·4815	2·4418	0·0682
	18·000	33·195	3·7603	2·4089	1·5610	2·3250	0·0910
	20·000	35·449	4·2404	2·5761	1·6460	2·2068	0·1170
	22·000	37·816	4·7587	2·7398	1·7369	2·0871	0·1459
	24·000	40·310	5·3167	2·8993	1·8338	1·9655	0·1774
	26·000	42·953	5·9169	3·0540	1·9374	1·8416	0·2114
	28·000	45·783	6·5636	3·2042	2·0484	1·7142	0·2478
	30·000	48·865	7·2659	3·3507	2·1685	1·5815	0·2868
	32·000	52·333	8·0435	3·4957	2·3010	1·4392	0·3290
	34·000	56·535	8·9522	3·6460	2·4554	1·2763	0·3769
	36·000	64·193	10·4534	3·8579	2·7096	1·0052	0·4527
	36·054	65·522	10·6870	3·8874	2·7491	0·9616	0·4641

TABLE V

FLOW OF DRY AIR THROUGH A PLANE OBLIQUE SHOCK WAVE

M_1	δ	β	p_2/p_1	ρ_2/ρ_1	T_2/T_1	M_2	$\Delta S/c_V$
3·35	36·000	66·804	10·9039	3·9141	2·7858	0·9204	0·4746
	34·000	72·784	11·7889	4·0162	2·9353	0·7424	0·5165
	32·000	75·330	12·0964	4·0494	2·9872	0·6748	0·5307
	30·000	77·167	12·2904	4·0697	3·0200	0·6298	0·5396
	28·000	78·646	12·4290	4·0840	3·0434	0·5962	0·5459
	26·000	79·903	12·5341	4·0946	3·0611	0·5698	0·5507
	24·000	81·009	12·6168	4·1030	3·0751	0·5483	0·5544
	22·000	82·004	12·6833	4·1096	3·0863	0·5306	0·5574
	20·000	82·916	12·7376	4·1150	3·0954	0·5157	0·5598
	18·000	83·763	12·7822	4·1194	3·1030	0·5033	0·5618
	16·000	84·559	12·8191	4·1230	3·1092	0·4928	0·5635
	14·000	85·314	12·8494	4·1260	3·1143	0·4841	0·5648
	12·000	86·036	12·8743	4·1284	3·1185	0·4768	0·5659
	10·000	86·733	12·8943	4·1303	3·1219	0·4709	0·5668
	8·000	87·409	12·9101	4·1319	3·1245	0·4662	0·5675
	6·000	88·070	12·9220	4·1330	3·1265	0·4626	0·5680
	4·000	88·719	12·9303	4·1338	3·1279	0·4601	0·5685
	2·000	89·361	12·9353	4·1343	3·1288	0·4586	0·5686
3·40	2·000	18·469	1·1870	1·1298	1·0506	3·2808	0·0002
	4·000	19·932	1·4011	1·2703	1·1030	3·1651	0·0016
	6·000	21·495	1·6446	1·4205	1·1578	3·0510	0·0050
	8·000	23·156	1·9196	1·5790	1·2157	2·9374	0·0113
	10·000	24·914	2·2278	1·7439	1·2774	2·8234	0·0208
	12·000	26·770	2·5707	1·9135	1·3435	2·7085	0·0338
	14·000	28·721	2·9495	2·0856	1·4142	2·5924	0·0504
	16·000	30·770	3·3652	2·2584	1·4901	2·4750	0·0706
	18·000	32·917	3·8185	2·4300	1·5714	2·3563	0·0941
	20·000	35·167	4·3102	2·5991	1·6583	2·2362	0·1209
	22·000	37·528	4·8412	2·7644	1·7512	2·1147	0·1506
	24·000	40·013	5·4127	2·9251	1·8504	1·9916	0·1829
	26·000	42·643	6·0270	3·0808	1·9563	1·8663	0·2176
	28·000	45·453	6·6883	3·2314	2·0698	1·7378	0·2548
	30·000	48·503	7·4049	3·3778	2·1922	1·6044	0·2943
	32·000	51·915	8·1951	3·5221	2·3268	1·4623	0·3371
	34·000	55·990	9·1078	3·6699	2·4817	1·3019	0·3850
	36·000	62·335	10·4210	3·8537	2·7042	1·0711	0·4511
	36·304	65·564	11·0210	3·9282	2·8056	0·9625	0·4802
	36·000	68·535	11·5235	3·9867	2·8905	0·8680	0·5041
	34·000	73·195	12·2027	4·0606	3·0052	0·7317	0·5356
	32·000	75·606	12·4969	4·0909	3·0548	0·6678	0·5490
	30·000	77·380	12·6866	4·1099	3·0868	0·6244	0·5576
	28·000	78·819	12·8234	4·1234	3·1099	0·5917	0·5637
	26·000	80·049	12·9279	4·1336	3·1275	0·5659	0·5683
	24·000	81·133	13·0102	4·1415	3·1414	0·5449	0·5720
	22·000	82·111	13·0767	4·1479	3·1526	0·5274	0·5749
	20·000	83·008	13·1309	4·1530	3·1618	0·5129	0·5773
	18·000	83·842	13·1757	4·1573	3·1693	0·5006	0·5793
	16·000	84·627	13·2126	4·1607	3·1755	0·4903	0·5809
	14·000	85·371	13·2431	4·1636	3·1807	0·4816	0·5823
	12·000	86·084	13·2680	4·1659	3·1849	0·4744	0·5834
	10·000	86·772	13·2882	4·1678	3·1883	0·4686	0·5843
	8·000	87·440	13·3041	4·1693	3·1910	0·4639	0·5850

TABLE V

FLOW OF DRY AIR THROUGH A PLANE OBLIQUE SHOCK WAVE

M_1	δ	β	p_2/p_1	ρ_2/ρ_1	T_2/T_1	M_2	$\Delta S/c_V$
3·40	6·000	88·092	13·3160	4·1704	3·1930	0·4604	0·5855
	4·000	88·734	13·3244	4·1712	3·1944	0·4579	0·5858
	2·000	89·369	13·3293	4·1716	3·1952	0·4564	0·5861
3·45	2·000	18·211	1·1897	1·1316	1·0513	3·3286	0·0002
	4·000	19·672	1·4073	1·2742	1·1044	3·2106	0·0016
	6·000	21·233	1·6552	1·4268	1·1600	3·0945	0·0052
	8·000	22·893	1·9355	1·5878	1·2190	2·9787	0·0116
	10·000	24·651	2·2501	1·7554	1·2818	2·8626	0·0215
	12·000	26·506	2·6005	1·9276	1·3491	2·7456	0·0350
	14·000	28·457	2·9879	2·1022	1·4213	2·6274	0·0522
	16·000	30·504	3·4133	2·2774	1·4988	2·5079	0·0730
	18·000	32·649	3·8775	2·4512	1·5819	2·3872	0·0973
	20·000	34·896	4·3811	2·6222	1·6708	2·2653	0·1248
	22·000	37·252	4·9249	2·7890	1·7658	2·1420	0·1553
	24·000	39·729	5·5101	2·9509	1·8673	2·0173	0·1884
	26·000	42·347	6·1389	3·1074	1·9756	1·8905	0·2240
	28·000	45·138	6·8152	3·2585	2·0915	1·7610	0·2618
	30·000	48·160	7·5467	3·4050	2·2164	1·6268	0·3021
	32·000	51·523	8·3502	3·5486	2·3531	1·4847	0·3453
	34·000	55·490	9·2697	3·6943	2·5092	1·3261	0·3934
	36·000	61·226	10·5106	3·8652	2·7193	1·1132	0·4555
	36·544	65·606	11·3602	3·9680	2·8629	0·9634	0·4964
	36·000	69·522	12·0299	4·0423	2·9760	0·8392	0·5276
	34·000	73·566	12·6185	4·1031	3·0753	0·7220	0·5545
	32·000	75·863	12·9018	4·1311	3·1231	0·6613	0·5672
	30·000	77·579	13·0879	4·1489	3·1545	0·6194	0·5755
	28·000	78·982	13·2233	4·1617	3·1773	0·5875	0·5814
	26·000	80·186	'13·3271	4·1714	3·1948	0·5622	0·5859
	24·000	81·250	13·4093	4·1790	3·2087	0·5416	0·5896
	22·000	82·211	13·4757	4·1851	3·2199	0·5245	0·5924
	20·000	83·095	13·5300	4·1901	3·2291	0·5101	0·5948
	18·000	83·917	13·5749	4·1942	3·2366	0·4980	0·5967
	16·000	84·691	13·6119	4·1975	3·2429	0·4878	0·5984
	14·000	85·426	13·6425	4·2003	3·2480	0·4793	0·5997
	12·000	86·130	13·6676	4·2025	3·2522	0·4722	0·6008
	10·000	86·809	13·6879	4·2043	3·2557	0·4664	0·6016
	8·000	87·469	13·7038	4·2058	3·2584	0·4618	0·6023
	6·000	88·114	13·7159	4·2068	3·2604	0·4583	0·6029
	4·000	88·749	13·7243	4·2076	3·2618	0·4558	0·6032
	2·000	89·376	13·7293	4·2080	3·2626	0·4543	0·6034
3·50	2·000	17·960	1·1924	1·1334	1·0520	3·3763	0·0002
	4·000	19·419	1·4135	1·2782	1·1058	3·2561	0·0017
	6·000	20·979	1·6658	1·4331	1·1623	3·1378	0·0054
	8·000	22·638	1·9516	1·5967	1·2223	3·0199	0·0121
	10·000	24·396	2·2727	1·7669	1·2862	2·9017	0·0224
	12·000	26·251	2·6307	1·9417	1·3548	2·7825	0·0363
	14·000	28·201	3·0268	2·1189	1·4285	2·6621	0·0540
	16·000	30·248	3·4621	2·2965	1·5076	2·5406	0·0755
	18·000	32·391	3·9373	2·4724	1·5925	2·4179	0·1005
	20·000	34·635	4·4529	2·6452	1·6834	2·2940	0·1288
	22·000	36·986	5·0098	2·8136	1·7806	2·1689	0·1601

TABLE V

FLOW OF DRY AIR THROUGH A PLANE OBLIQUE SHOCK WAVE

M_1	δ	β	p_2/p_1	ρ_2/ρ_1	T_2/T_1	M_2	$\Delta S/c_v$
3.50	24.000	39.456	5.6091	2.9767	1.8843	2.0425	0.1940
	26.000	42.063	6.2526	3.1340	1.9951	1.9144	0.2303
	28.000	44.838	6.9442	3.2856	2.1135	1.7836	0.2690
	30.000	47.835	7.6910	3.4320	2.2410	1.6487	0.3100
	32.000	51.153	8.5088	3.5750	2.3801	1.5064	0.3537
	34.000	55.027	9.4371	3.7189	2.5376	1.3492	0.4019
	36.000	60.364	10.6391	3.8814	2.7410	1.1476	0.4618
	36.776	65.647	11.7044	4.0069	2.9211	0.9643	0.5126
	36.000	70.266	12.5058	4.0918	3.0563	0.8180	0.5494
	34.000	73.904	13.0372	4.1441	3.1460	0.7132	0.5732
	32.000	76.101	13.3113	4.1700	3.1922	0.6553	0.5852
	30.000	77.766	13.4944	4.1868	3.2230	0.6146	0.5932
	28.000	79.136	13.6286	4.1990	3.2457	0.5836	0.5991
	26.000	80.315	13.7319	4.2083	3.2631	0.5588	0.6036
	24.000	81.361	13.8139	4.2155	3.2769	0.5385	0.6071
	22.000	82.307	13.8804	4.2214	3.2881	0.5216	0.6099
	20.000	83.177	13.9348	4.2262	3.2973	0.5075	0.6122
	18.000	83.988	13.9798	4.2301	3.3049	0.4955	0.6142
	16.000	84.752	14.0170	4.2333	3.3111	0.4855	0.6157
	14.000	85.478	14.0478	4.2360	3.3163	0.4770	0.6171
	12.000	86.173	14.0730	4.2381	3.3206	0.4700	0.6181
	10.000	86.845	14.0934	4.2399	3.3240	0.4643	0.6190
	8.000	87.497	14.1094	4.2413	3.3267	0.4597	0.6197
	6.000	88.135	14.1216	4.2423	3.3288	0.4563	0.6202
	4.000	88.762	14.1300	4.2430	3.3302	0.4538	0.6206
	2.000	89.383	14.1350	4.2435	3.3310	0.4524	0.6208
3.55	2.000	17.717	1.1951	1.1353	1.0527	3.4239	0.0002
	4.000	19.174	1.4197	1.2822	1.1073	3.3016	0.0017
	6.000	20.733	1.6765	1.4395	1.1647	3.1811	0.0056
	8.000	22.392	1.9678	1.6056	1.2256	3.0610	0.0126
	10.000	24.149	2.2954	1.7785	1.2907	2.9405	0.0231
	12.000	26.004	2.6611	1.9559	1.3605	2.8192	0.0376
	14.000	27.955	3.0662	2.1357	1.4357	2.6967	0.0559
	16.000	30.001	3.5115	2.3156	1.5165	2.5730	0.0780
	18.000	32.143	3.9978	2.4936	1.6032	2.4482	0.1038
	20.000	34.384	4.5257	2.6682	1.6961	2.3223	0.1328
	22.000	36.731	5.0959	2.8381	1.7955	2.1954	0.1649
	24.000	39.195	5.7095	3.0024	1.9017	2.0674	0.1997
	26.000	41.792	6.3681	3.1605	2.0149	1.9377	0.2369
	28.000	44.552	7.0753	3.3125	2.1360	1.8058	0.2762
	30.000	47.525	7.8380	3.4590	2.2660	1.6700	0.3179
	32.000	50.804	8.6707	3.6014	2.4076	1.5274	0.3622
	34.000	54.597	9.6095	3.7438	2.5668	1.3713	0.4107
	36.000	59.643	10.7895	3.9001	2.7665	1.1775	0.4691
	36.999	65.687	12.0538	4.0448	2.9801	0.9651	0.5287
	36.000	70.874	12.9685	4.1375	3.1344	0.8008	0.5701
	34.000	74.214	13.4592	4.1836	3.2171	0.7051	0.5917
	32.000	76.324	13.7256	4.2077	3.2620	0.6496	0.6033
	30.000	77.942	13.9061	4.2236	3.2924	0.6101	0.6110
	28.000	79.281	14.0393	4.2352	3.3149	0.5798	0.6167
	26.000	80.438	14.1423	4.2441	3.3322	0.5555	0.6211
	24.000	81.466	14.2242	4.2511	3.3461	0.5355	0.6245

TABLE V

FLOW OF DRY AIR THROUGH A PLANE OBLIQUE SHOCK WAVE

M_1	δ	β	p_2/p_1	ρ_2/ρ_1	T_2/T_1	M_2	$\Delta S/c_V$
3·55	22·000	82·398	14·2908	4·2567	3·3573	0·5189	0·6274
	20·000	83·256	14·3454	4·2613	3·3665	0·5049	0·6297
	18·000	84·056	14·3905	4·2650	3·3741	0·4931	0·6316
	16·000	84·810	14·4279	4·2681	3·3804	0·4832	0·6332
	14·000	85·527	14·4588	4·2707	3·3856	0·4749	0·6345
	12·000	86·215	14·4842	4·2728	3·3898	0·4679	0·6355
	10·000	86·879	14·5047	4·2745	3·3933	0·4623	0·6364
	8·000	87·524	14·5209	4·2758	3·3960	0·4577	0·6370
	6·000	88·155	14·5331	4·2768	3·3981	0·4543	0·6376
	4·000	88·775	14·5416	4·2775	3·3995	0·4519	0·6379
	2·000	89·389	14·5466	4·2780	3·4004	0·4505	0·6381
3·60	2·000	17·481	1·1978	1·1371	1·0534	3·4715	0·0002
	4·000	18·936	1·4260	1·2861	1·1088	3·3469	0·0018
	6·000	20·494	1·6873	1·4459	1·1670	3·2242	0·0058
	8·000	22·153	1·9841	1·6146	1·2289	3·1019	0·0131
	10·000	23·911	2·3184	1·7901	1·2951	2·9792	0·0240
	12·000	25·766	2·6919	1·9702	1·3663	2·8556	0·0388
	14·000	27·717	3·1060	2·1525	1·4430	2·7309	0·0578
	16·000	29·762	3·5615	2·3347	1·5255	2·6051	0·0806
	18·000	31·904	4·0591	2·5149	1·6141	2·4782	0·1071
	20·000	34·143	4·5995	2·6913	1·7091	2·3504	0·1370
	22·000	36·486	5·1833	2·8626	1·8107	2·2216	0·1699
	24·000	38·944	5·8113	3·0280	1·9192	2·0918	0·2054
	26·000	41·532	6·4853	3·1869	2·0350	1·9607	0·2434
	28·000	44·279	7·2086	3·3393	2·1587	1·8275	0·2836
	30·000	47·230	7·9875	3·4858	2·2914	1·6908	0·3259
	32·000	50·473	8·8358	3·6278	2·4356	1·5479	0·3709
	34·000	54·196	9·7865	3·7687	2·5968	1·3925	0·4196
	36·000	59·015	10·9549	3·9203	2·7944	1·2046	0·4771
	37·213	65·727	12·4083	4·0819	3·0399	0·9660	0·5450
	36·000	71·391	13·4248	4·1805	3·2113	0·7864	0·5902
	34·000	74·499	13·8848	4·2218	3·2889	0·6977	0·6101
	32·000	76·532	14·1448	4·2443	3·3327	0·6443	0·6212
	30·000	78·107	14·3231	4·2594	3·3627	0·6059	0·6287
	28·000	79·419	14·4555	4·2704	3·3850	0·5762	0·6343
	26·000	80·555	14·5582	4·2789	3·4023	0·5523	0·6386
	24·000	81·566	14·6402	4·2856	3·4161	0·5327	0·6420
	22·000	82·484	14·7069	4·2910	3·4274	0·5163	0·6448
	20·000	83·330	14·7616	4·2954	3·4366	0·5025	0·6471
	18·000	84·121	14·8070	4·2991	3·4442	0·4909	0·6490
	16·000	84·866	14·8446	4·3021	3·4506	0·4810	0·6505
	14·000	85·574	14·8757	4·3046	3·4558	0·4728	0·6518
	12·000	86·254	14·9012	4·3066	3·4601	0·4659	0·6528
	10·000	86·911	14·9218	4·3082	3·4636	0·4603	0·6537
	8·000	87·549	14·9381	4·3095	3·4663	0·4558	0·6544
	6·000	88·174	14·9504	4·3105	3·4684	0·4524	0·6549
	4·000	88·788	14·9590	4·3112	3·4698	0·4501	0·6552
	2·000	89·396	14·9641	4·3116	3·4707	0·4486	0·6555
3·65	2·000	17·251	1·2005	1·1389	1·0541	3·5191	0·0002
	4·000	18·705	1·4323	1·2901	1·1102	3·3922	0·0019
	6·000	20·263	1·6982	1·4523	1·1693	3·2672	0·0061

TABLE V

FLOW OF DRY AIR THROUGH A PLANE OBLIQUE SHOCK WAVE

M_1	δ	β	p_2/p_1	ρ_2/ρ_1	T_2/T_1	M_2	$\Delta S/c_v$
3·65	8·000	21·921	2·0006	1·6236	1·2322	3·1427	0·0135
	10·000	23·680	2·3416	1·8018	1·2996	3·0178	0·0248
	12·000	25·535	2·7231	1·9845	1·3722	2·8919	0·0402
	14·000	27·487	3·1462	2·1693	1·4503	2·7649	0·0597
	16·000	29·532	3·6120	2·3539	1·5345	2·6369	0·0832
	18·000	31·673	4·1212	2·5361	1·6250	2·5079	0·1105
	20·000	33·910	4·6743	2·7143	1·7221	2·3780	0·1412
	22·000	36·251	5·2718	2·8870	1·8260	2·2474	0·1745
	24·000	38·703	5·9146	3·0535	1·9370	2·1159	0·2113
	26·000	41·283	6·6043	3·2132	2·0554	1·9833	0·2501
	28·000	44·017	7·3439	3·3660	2·1818	1·8488	0·2910
	30·000	46·949	8·1395	3·5125	2·3173	1·7111	0·3341
	32·000	50·159	9·0040	3·6540	2·4641	1·5677	0·3797
	34·000	53·820	9·9679	3·7937	2·6275	1·4130	0·4287
	36·000	58·457	11·1316	3·9414	2·8243	1·2294	0·4855
	37·420	65·766	12·7680	4·1180	3·1006	0·9668	0·5612
	36·000	71·843	13·8784	4·2212	3·2878	0·7738	0·6099
	34·000	74·763	14·3145	4·2587	3·3613	0·6908	0·6284
	32·000	76·727	14·5690	4·2798	3·4041	0·6393	0·6391
	30·000	78·264	14·7454	4·2941	3·4339	0·6018	0·6464
	28·000	79·549	14·8771	4·3047	3·4560	0·5727	0·6518
	26·000	80·665	14·9798	4·3128	3·4733	0·5493	0·6561
	24·000	81·661	15·0618	4·3192	3·4871	0·5299	0·6595
	22·000	82·566	15·1287	4·3244	3·4984	0·5138	0·6622
	20·000	83·402	15·1837	4·3287	3·5077	0·5002	0·6644
	18·000	84·182	15·2292	4·3322	3·5153	0·4887	0·6663
	16·000	84·918	15·2670	4·3351	3·5217	0·4790	0·6679
	14·000	85·619	15·2983	4·3375	3·5270	0·4708	0·6691
	12·000	86·292	15·3240	4·3395	3·5313	0·4640	0·6701
	10·000	86·942	15·3448	4·3411	3·5348	0·4584	0·6710
	8·000	87·573	15·3612	4·3423	3·5376	0·4540	0·6717
	6·000	88·192	15·3736	4·3433	3·5396	0·4506	0·6722
	4·000	88·800	15·3822	4·3439	3·5411	0·4483	0·6725
	2·000	89·401	15·3874	4·3443	3·5420	0·4469	0·6727
3·70	2·000	17·029	1·2033	1·1408	1·0548	3·5667	0·0002
	4·000	18·481	1·4387	1·2941	1·1117	3·4375	0·0020
	6·000	20·038	1·7091	1·4587	1·1717	3·3102	0·0063
	8·000	21·697	2·0172	1·6326	1·2356	3·1834	0·0140
	10·000	23·456	2·3650	1·8135	1·3042	3·0561	0·0257
	12·000	25·312	2·7545	1·9989	1·3780	2·9279	0·0416
	14·000	27·264	3·1869	2·1862	1·4578	2·7987	0·0617
	16·000	29·310	3·6632	2·3731	1·5437	2·6684	0·0859
	18·000	31·450	4·1841	2·5573	1·6361	2·5373	0·1139
	20·000	33·686	4·7501	2·7372	1·7353	2·4054	0·1454
	22·000	36·024	5·3616	2·9114	1·8416	2·2729	0·1799
	24·000	38·471	6·0194	3·0789	1·9550	2·1397	0·2172
	26·000	41·044	6·7250	3·2393	2·0761	2·0055	0·2568
	28·000	43·767	7·4812	3·3925	2·2052	1·8697	0·2985
	30·000	46·681	8·2939	3·5390	2·3436	1·7310	0·3423
	32·000	49·862	9·1752	3·6801	2·4932	1·5871	0·3885
	34·000	53·467	10·1534	3·8187	2·6589	1·4327	0·4379
	36·000	57·953	11·3174	3·9631	2·8557	1·2526	0·4943

TABLE V

FLOW OF DRY AIR THROUGH A PLANE OBLIQUE SHOCK WAVE

M_1	δ	β	p_2/p_1	ρ_2/ρ_1	T_2/T_1	M_2	$\Delta S/c_V$
3·70	37·619	65·803	13·1327	4·1532	3·1621	0·9675	0·5774
	36·000	72·245	14·3316	4·2601	3·3642	0·7628	0·6291
	34·000	75·008	14·7484	4·2944	3·4344	0·6844	0·6465
	32·000	76·910	14·9983	4·3143	3·4764	0·6346	0·6569
	30·000	78·412	15·1731	4·3279	3·5059	0·5980	0·6640
	28·000	79·672	15·3044	4·3386	3·5280	0·5695	0·6694
	26·000	80·770	15·4069	4·3458	3·5453	0·5464	0·6735
	24·000	81·751	15·4891	4·3520	3·5591	0·5273	0·6769
	22·000	82·644	15·5562	4·3570	3·5704	0·5114	0·6795
	20·000	83·469	15·6114	4·3611	3·5797	0·4979	0·6817
	18·000	84·241	15·6572	4·3645	3·5874	0·4866	0·6836
	16·000	84·969	15·6953	4·3673	3·5938	0·4770	0·6851
	14·000	85·662	15·7268	4·3696	3·5991	0·4689	0·6864
	12·000	86·328	15·7526	4·3715	3·6035	0·4621	0·6874
	10·000	86·971	15·7736	4·3731	3·6070	0·4566	0·6883
	8·000	87·597	15·7901	4·3743	3·6098	0·4522	0·6889
	6·000	88·209	15·8026	4·3752	3·6119	0·4489	0·6894
	4·000	88·811	15·8113	4·3758	3·6133	0·4466	0·6898
	2·000	89·407	15·8165	4·3762	3·6142	0·4452	0·6900
3·75	2·000	16·812	1·2060	1·1426	1·0555	3·6142	0·0003
	4·000	18·264	1·4450	1·2981	1·1132	3·4826	0·0020
	6·000	19·820	1·7201	1·4652	1·1740	3·3530	0·0065
	8·000	21·479	2·0339	1·6416	1·2389	3·2239	0·0145
	10·000	23·239	2·3887	1·8252	1·3087	3·0943	0·0266
	12·000	25·096	2·7863	2·0132	1·3840	2·9638	0·0430
	14·000	27·049	3·2281	2·2031	1·4652	2·8322	0·0637
	16·000	29·095	3·7150	2·3923	1·5529	2·6997	0·0886
	18·000	31·235	4·2478	2·5786	1·6473	2·5664	0·1174
	20·000	33·470	4·8268	2·7602	1·7487	2·4324	0·1497
	22·000	35·805	5·4525	2·9357	1·8573	2·2980	0·1851
	24·000	38·248	6·1256	3·1043	1·9733	2·1630	0·2232
	26·000	40·814	6·8474	3·2653	2·0970	2·0273	0·2636
	28·000	43·527	7·6206	3·4189	2·2290	1·8901	0·3061
	30·000	46·424	8·4508	3·5654	2·3702	1·7504	0·3507
	32·000	49·578	9·3493	3·7061	2·5227	1·6059	0·3974
	34·000	53·134	10·3428	3·8436	2·6909	1·4517	0·4473
	36·000	57·493	11·5108	3·9852	2·8884	1·2744	0·5035
	37·811	65·840	13·5025	4·1876	3·2244	0·9683	0·5936
	36·000	72·606	14·7858	4·2974	3·4407	0·7528	0·6481
	34·000	75·236	15·1868	4·3290	3·5082	0·6785	0·6645
	32·000	77·083	15·4327	4·3477	3·5496	0·6301	0·6746
	30·000	78·551	15·6062	4·3607	3·5788	0·5944	0·6816
	28·000	79·789	15·7371	4·3704	3·6008	0·5664	0·6868
	26·000	80·870	15·8397	4·3779	3·6181	0·5436	0·6909
	24·000	81·837	15·9221	4·3838	3·6320	0·5249	0·6942
	22·000	82·719	15·9894	4·3887	3·6433	0·5091	0·6969
	20·000	83·534	16·0449	4·3927	3·6527	0·4958	0·6991
	18·000	84·297	16·0910	4·3959	3·6604	0·4845	0·7009
	16·000	85·017	16·1293	4·3987	3·6669	0·4750	0·7024
	14·000	85·703	16·1610	4·4009	3·6722	0·4670	0·7036
	12·000	86·362	16·1871	4·4027	3·6766	0·4604	0·7047
	10·000	86·999	16·2082	4·4042	3·6801	0·4549	0·7055

TABLE V

FLOW OF DRY AIR THROUGH A PLANE OBLIQUE SHOCK WAVE

M_1	δ	β	p_2/p_1	ρ_2/ρ_1	T_2/T_1	M_2	$\Delta S/c_V$
3·75	8·000	87·619	16·2248	4·4054	3·6829	0·4506	0·7061
	6·000	88·225	16·2374	4·4063	3·6851	0·4472	0·7066
	4·000	88·822	16·2462	4·4069	3·6866	0·4449	0·7070
	2·000	89·413	16·2514	4·4073	3·6874	0·4435	0·7072
3·80	2·000	16·602	1·2088	1·1445	1·0562	3·6617	0·0003
	4·000	18·052	1·4514	1·3022	1·1146	3·5277	0·0021
	6·000	19·608	1·7312	1·4716	1·1764	3·3957	0·0067
	8·000	21·268	2·0508	1·6507	1·2423	3·2642	0·0150
	10·000	23·028	2·4125	1·8370	1·3133	3·1323	0·0275
	12·000	24·887	2·8184	2·0277	1·3900	2·9994	0·0444
	14·000	26·840	3·2697	2·2200	1·4728	2·8655	0·0658
	16·000	28·887	3·7674	2·4115	1·5623	2·7307	0·0914
	18·000	31·027	4·3122	2·5998	1·6587	2·5952	0·1210
	20·000	33·262	4·9045	2·7831	1·7623	2·4592	0·1541
	22·000	35·594	5·5446	2·9600	1·8732	2·3228	0·1903
	24·000	38·034	6·2332	3·1295	1·9918	2·1860	0·2293
	26·000	40·594	6·9715	3·2912	2·1182	2·0487	0·2705
	28·000	43·296	7·7620	3·4451	2·2531	1·9102	0·3138
	30·000	46·179	8·6100	3·5916	2·3973	1·7694	0·3591
	32·000	49·308	9·5263	3·7319	2·5527	1·6242	0·4064
	34·000	52·820	10·5359	3·8684	2·7236	1·4701	0·4567
	36·000	57·069	11·7109	4·0076	2·9222	1·2950	0·5129
	37·997	65·877	13·8774	4·2211	3·2876	0·9690	0·6098
	36·000	72·935	15·2418	4·3332	3·5175	0·7439	0·6668
	34·000	75·450	15·6297	4·3625	3·5828	0·6729	0·6825
	32·000	77·246	15·8722	4·3802	3·6236	0·6259	0·6922
	30·000	78·684	16·0448	4·3926	3·6526	0·5909	0·6990
	28·000	79·900	16·1754	4·4019	3·6746	0·5634	0·7042
	26·000	80·965	16·2781	4·4091	3·6919	0·5410	0·7082
	24·000	81·919	16·3608	4·4149	3·7058	0·5225	0·7115
	22·000	82·790	16·4284	4·4195	3·7172	0·5069	0·7141
	20·000	83·596	16·4842	4·4234	3·7266	0·4937	0·7163
	18·000	84·350	16·5305	4·4266	3·7344	0·4826	0·7181
	16·000	85·063	16·5691	4·4292	3·7409	0·4732	0·7196
	14·000	85·742	16·6010	4·4314	3·7463	0·4652	0·7208
	12·000	86·395	16·6273	4·4331	3·7507	0·4586	0·7218
	10·000	87·026	16·6486	4·4346	3·7543	0·4532	0·7226
	8·000	87·640	16·6654	4·4357	3·7571	0·4489	0·7233
	6·000	88·241	16·6781	4·4366	3·7592	0·4456	0·7238
	4·000	88·832	16·6870	4·4372	3·7607	0·4433	0·7241
	2·000	89·418	16·6922	4·4375	3·7616	0·4420	0·7243
3·85	2·000	16·397	1·2115	1·1463	1·0569	3·7091	0·0003
	4·000	17·846	1·4579	1·3062	1·1161	3·5727	0·0022
	6·000	19·402	1·7423	1·4781	1·1787	3·4384	0·0069
	8·000	21·062	2·0678	1·6599	1·2457	3·3045	0·0155
	10·000	22·824	2·4366	1·8488	1·3179	3·1701	0·0284
	12·000	24·684	2·8508	2·0421	1·3960	3·0349	0·0458
	14·000	26·638	3·3117	2·2370	1·4804	2·8986	0·0679
	16·000	28·686	3·8204	2·4307	1·5717	2·7615	0·0942
	18·000	30·827	4·3774	2·6210	1·6701	2·6237	0·1246
	20·000	33·060	4·9831	2·8059	1·7759	2·4856	0·1585
	22·000	35·391	5·6379	2·9841	1·8893	2·3472	0·1956

TABLE V

FLOW OF DRY AIR THROUGH A PLANE OBLIQUE SHOCK WAVE

M_1	δ	β	p_2/p_1	ρ_2/ρ_1	T_2/T_1	M_2	$\Delta S/c_v$
3·85	24·000	37·827	6·3423	3·1546	2·0105	2·2087	0·2354
	26·000	40·382	7·0973	3·3170	2·1397	2·0698	0·2774
	28·000	43·075	7·9054	3·4711	2·2775	1·9299	0·3216
	30·000	45·944	8·7716	3·6176	2·4247	1·7880	0·3675
	32·000	49·051	9·7062	3·7575	2·5832	1·6421	0·4156
	34·000	52·522	10·7327	3·8934	2·7569	1·4879	0·4664
	36·000	56·677	11·9170	4·0302	2·9570	1·3147	0·5224
	38·000	63·503	13·6954	4·2050	3·2569	1·0545	0·6020
	38·175	65·912	14·2575	4·2539	3·3517	0·9697	0·6260
	38·000	68·166	14·7465	4·2942	3·4340	0·8937	0·6465
	36·000	73·235	15·7005	4·3677	3·5947	0·7357	0·6854
	34·000	75·650	16·0774	4·3950	3·6581	0·6677	0·7003
	32·000	77·401	16·3171	4·4118	3·6985	0·6219	0·7097
	30·000	78·810	16·4888	4·4237	3·7274	0·5876	0·7164
	28·000	80·006	16·6193	4·4326	3·7493	0·5606	0·7215
	26·000	81·055	16·7222	4·4395	3·7667	0·5385	0·7255
	24·000	81·997	16·8052	4·4451	3·7806	0·5202	0·7287
	22·000	82·858	16·8731	4·4496	3·7920	0·5048	0·7313
	20·000	83·655	16·9292	4·4533	3·8015	0·4918	0·7335
	18·000	84·401	16·9759	4·4564	3·8093	0·4807	0·7352
	16·000	85·107	17·0147	4·4589	3·8159	0·4714	0·7367
	14·000	85·780	17·0469	4·4610	3·8213	0·4635	0·7380
	12·000	86·426	17·0734	4·4628	3·8257	0·4570	0·7390
	10·000	87·052	17·0948	4·4642	3·8294	0·4516	0·7398
	8·000	87·660	17·1118	4·4653	3·8322	0·4473	0·7404
	6·000	88·256	17·1246	4·4661	3·8344	0·4441	0·7409
	4·000	88·842	17·1336	4·4667	3·8359	0·4418	0·7412
	2·000	89·423	17·1389	4·4670	3·8368	0·4404	0·7415
3·90	2·000	16·197	1·2143	1·1482	1·0576	3·7566	0·0003
	4·000	17·646	1·4643	1·3103	1·1176	3·6177	0·0022
	6·000	19·202	1·7536	1·4847	1·1811	3·4809	0·0072
	8·000	20·863	2·0849	1·6690	1·2492	3·3446	0·0160
	10·000	22·626	2·4609	1·8607	1·3226	3·2078	0·0293
	12·000	24·487	2·8835	2·0566	1·4020	3·0701	0·0473
	14·000	26·443	3·3542	2·2540	1·4881	2·9314	0·0700
	16·000	28·492	3·8739	2·4499	1·5812	2·7919	0·0971
	18·000	30·632	4·4433	2·6422	1·6817	2·6519	0·1283
	20·000	32·866	5·0627	2·8287	1·7898	2·5116	0·1630
	22·000	35·195	5·7324	3·0082	1·9056	2·3713	0·2010
	24·000	37·628	6·4528	3·1796	2·0294	2·2310	0·2416
	26·000	40·177	7·2248	3·3425	2·1615	2·0905	0·2845
	28·000	42·863	8·0509	3·4970	2·3022	1·9492	0·3294
	30·000	45·719	8·9355	3·6434	2·4525	1·8062	0·3761
	32·000	48·805	9·8888	3·7829	2·6141	1·6596	0·4247
	34·000	52·239	10·9330	3·9176	2·7907	1·5052	0·4760
	36·000	56·311	12·1287	4·0528	2·9927	1·3334	0·5322
	38·000	62·509	13·8086	4·2151	3·2760	1·0931	0·6068
	38·348	65·947	14·6426	4·2858	3·4165	0·9704	0·6422
	38·000	69·076	15·3278	4·3398	3·5319	0·8653	0·6703
	36·000	73·511	16·1623	4·4010	3·6724	0·7282	0·7037
	34·000	75·838	16·5300	4·4265	3·7343	0·6627	0·7181
	32·000	77·547	16·7672	4·4426	3·7742	0·6181	0·7272
	30·000	78·929	16·9382	4·4539	3·8030	0·5845	0·7338

TABLE V

FLOW OF DRY AIR THROUGH A PLANE OBLIQUE SHOCK WAVE

M_1	δ	β	p_2/p_1	ρ_2/ρ_1	T_2/T_1	M_2	$\Delta S/c_v$
3·90	28·000	80·107	17·0688	4·4625	3·8250	0·5578	0·7388
	26·000	81·142	17·1719	4·4691	3·8423	0·5361	0·7427
	24·000	82·072	17·2552	4·4745	3·8563	0·5180	0·7459
	22·000	82·923	17·3235	4·4789	3·8678	0·5027	0·7485
	20·000	83·711	17·3800	4·4825	3·8773	0·4899	0·7506
	18·000	84·450	17·4270	4·4854	3·8852	0·4789	0·7524
	16·000	85·149	17·4661	4·4879	3·8918	0·4697	0·7538
	14·000	85·816	17·4986	4·4899	3·8973	0·4619	0·7550
	12·000	86·456	17·5253	4·4916	3·9018	0·4554	0·7560
	10·000	87·076	17·5469	4·4930	3·9054	0·4501	0·7569
	8·000	87·680	17·5640	4·4940	3·9083	0·4458	0·7575
	6·000	88·270	17·5769	4·4948	3·9105	0·4426	0·7580
	4·000	88·852	17·5860	4·4954	3·9120	0·4403	0·7583
	2·000	89·427	17·5913	4·4957	3·9129	0·4390	0·7585
3·95	2·000	16·003	1·2171	1·1500	1·0583	3·8039	0·0003
	4·000	17·451	1·4709	1·3143	1·1191	3·6625	0·0023
	6·000	19·007	1·7649	1·4912	1·1835	3·5233	0·0075
	8·000	20·669	2·1022	1·6782	1·2526	3·3845	0·0166
	10·000	22·433	2·4854	1·8726	1·3273	3·2453	0·0303
	12·000	24·296	2·9165	2·0712	1·4082	3·1051	0·0489
	14·000	26·253	3·3971	2·2710	1·4959	2·9640	0·0722
	16·000	28·303	3·9281	2·4692	1·5909	2·8221	0·1000
	18·000	30·445	4·5101	2·6633	1·6934	2·6798	0·1320
	20·000	32·678	5·1433	2·8514	1·8038	2·5374	0·1676
	22·000	35·006	5·8281	3·0321	1·9221	2·3951	0·2064
	24·000	37·437	6·5647	3·2045	2·0486	2·2530	0·2479
	26·000	39·981	7·3540	3·3680	2·1835	2·1108	0·2916
	28·000	42·659	8·1982	3·5226	2·3273	1·9682	0·3373
	30·000	45·503	9·1018	3·6690	2·4807	1·8241	0·3847
	32·000	48·570	10·0742	3·8081	2·6455	1·6766	0·4340
	34·000	51·971	11·1366	3·9420	2·8251	1·5220	0·4858
	36·000	55·969	12·3454	4·0754	3·0293	1·3513	0·5421
	38·000	61·760	13·9724	4·2294	3·3036	1·1235	0·6139
	38·515	65·980	15·0327	4·3170	3·4822	0·9711	0·6583
	38·000	69·744	15·8675	4·3799	3·6228	0·8449	0·6920
	36·000	73·767	16·6277	4·4332	3·7507	0·7212	0·7219
	34·000	76·016	16·9875	4·4571	3·8113	0·6581	0·7357
	32·000	77·685	17·2226	4·4724	3·8509	0·6145	0·7446
	30·000	79·043	17·3932	4·4833	3·8796	0·5815	0·7511
	28·000	80·203	17·5239	4·4915	3·9015	0·5552	0·7560
	26·000	81·224	17·6273	4·4980	3·9189	0·5337	0·7599
	24·000	82·143	17·7110	4·5032	3·9330	0·5159	0·7630
	22·000	82·985	17·7797	4·5074	3·9446	0·5008	0·7655
	20·000	83·765	17·8365	4·5109	3·9541	0·4880	0·7676
	18·000	84·497	17·8839	4·5137	3·9621	0·4772	0·7694
	16·000	85·189	17·9233	4·5161	3·9687	0·4680	0·7708
	14·000	85·850	17·9560	4·5181	3·9742	0·4603	0·7721
	12·000	86·485	17·9830	4·5197	3·9788	0·4539	0·7731
	10·000	87·100	18·0048	4·5211	3·9824	0·4486	0·7739
	8·000	87·698	18·0221	4·5221	3·9853	0·4444	0·7745
	6·000	88·284	18·0351	4·5229	3·9875	0·4412	0·7750
	4·000	88·861	18·0442	4·5234	3·9891	0·4389	0·7753
	2·000	89·432	18·0497	4·5237	3·9900	0·4376	0·7755